ANIMALS AND THEIR WORLD

ANIMALS AND THEIR WORLD

by Mary Parker Buckles

A Ridge Press Book ☀ Blandford Press

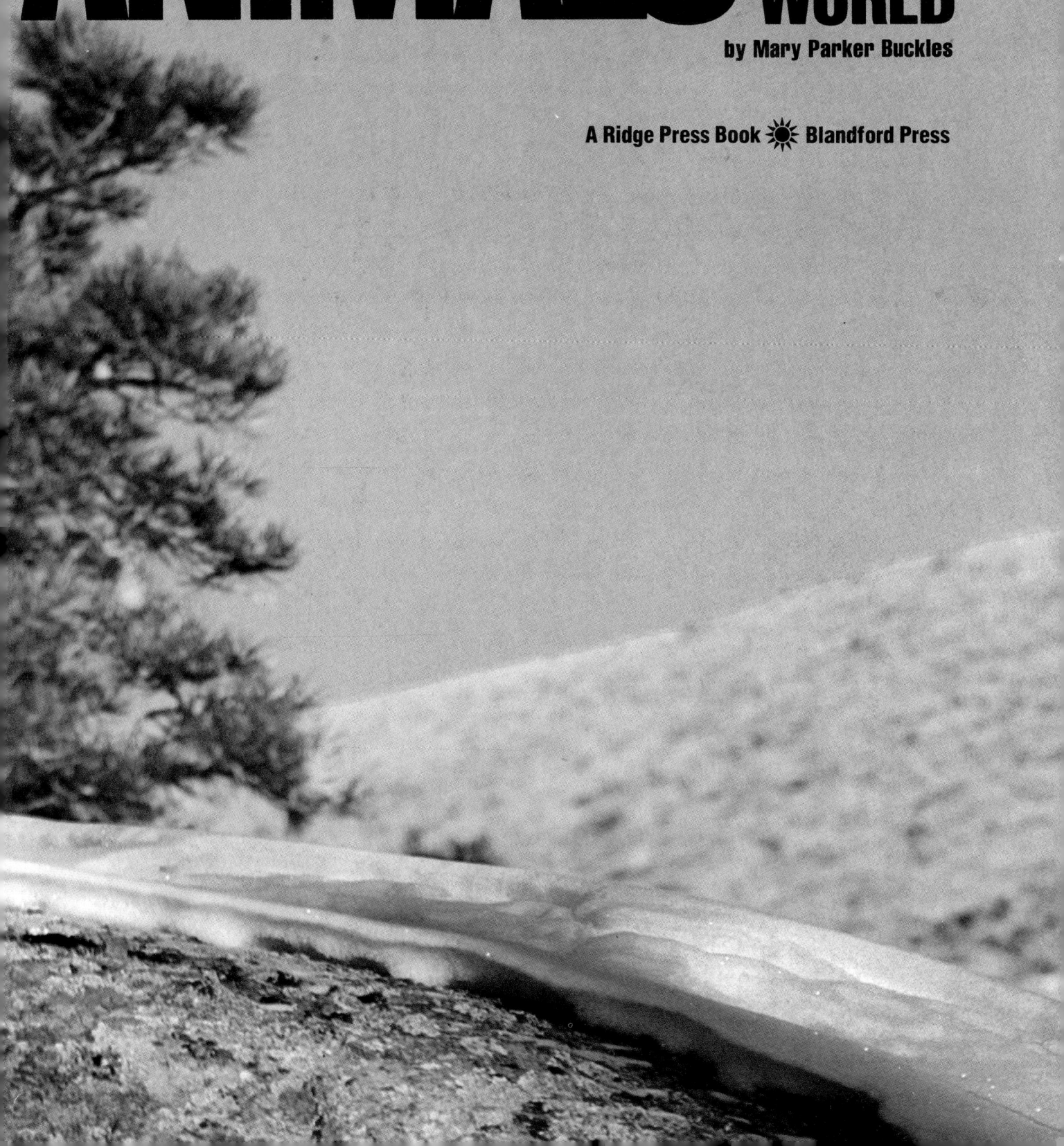

Editor-in-Chief: Jerry Mason
Editor: Adolph Suehsdorf
Art Director: Albert Squillace
Project Art Director: Harry Brocke
Associate Editor: Ronne Peltzman
Associate Editor: Joan Fisher
Art Associate: Nancy Mack
Art Associate: Liney Li
Art Production: Doris Mullane
Picture Editor: Marion Geisinger

First published in Great Britain
in 1979 by
Blandford Press Ltd.,
Link House, West Street,
Poole Dorset BH15 1LL

ISBN 0 7137 1027 6

Printed and bound in the Netherlands
by Smeets Offset, Weert.

Contents

Introduction

Animals and mammals—the two words conjure up similar visions of warm, furry creatures snuggled into domestic scenes, of stately elephants and giraffes in the grasslands, of tigers stalking prey in tropical forests. But although "animal" and "mammal" are often used interchangeably, the former is, in its strictest sense, a much broader category of living things than the latter. (Between 1 and 1.5 million animal species have been described—birds, insects, reptiles, mammals, and so forth. Only some 12,000 of these are living mammals.)

An animal is generally regarded as a mobile form of life that does not manufacture its own food, whereas a mammal is a warm-blooded, backboned animal with a body covering of hair or fur. It produces milk for its offspring, which are born live (without a hard shell), and it has other anatomical distinctions mentioned below. This book is about the world's wild mammals.

To grasp the true sense of the word "mammal," it's best to look first at its derivation. Taken from *mamma,* the Latin word for "breast," it connotes a whole world of nurturing that no other members of the animal kingdom can lay claim to. Birds come close, but they cannot match the more highly evolved mammals' abilities to parent and be parented. Infancy is a very special time in a mammal's life. Food is easily available then from the mother, and a great deal of cuddling and grooming, some transportation, and the simple reassurance of physical closeness often accompany feeding. In some species this kind of parental nurturing continues well after weaning is completed, and leads to instruction in food-gathering, self-defense, and other aspects of survival. The capacity to learn these necessary skills is an indication of the mammals' highly developed brains and remarkable intelligence.

Part of the vitality characteristic of mammals derives from their warm-bloodedness. Mammals maintain a constant body temperature (between 95 and 100.4° F [35–38° C]) that is relatively independent of the external environment. Temperature is maintained through metabolic activity, through the help of insulating "coats" of hair (or, in most marine mammals, layers of blubber), and through mechanisms such as sweating and shivering. It's as if mammals carry their heating and cooling systems around with them, and are therefore freed of having to rely on external sources of warmth or cold. Because of this, mammals can live and reproduce in many different climates.

Other, finer, points help round out the definition of a mammal. A four-chambered heart with a uniquely positioned major artery keeps oxygen-depleted blood separate from oxygen-rich blood, and thereby contributes to the efficient absorption of oxygen into a mammal's body tissues. (Specialized red blood cells help further.) The body cavities of mammals are separated into two parts by a muscle sheet known as the diaphragm. Along with some other muscles, this sheet regulates the size of the thoracic cavity and thus controls air flow through the lungs. There are specialized bone structures in the lower jaw and the middle ear, and, in all but a few species, seven vertebrae in the neck.

Adaptations

A mammal's adaptations are its means of coping with its enemies, its physical habitat, various climatic conditions, and many, many other environmental factors. Adaptations may be structural (the marsupials' pouch, for instance), functional (the springhare's use of its clawed front feet for digging), behavioral (the gorilla's self-defense through various threat displays), or any combination of these. Whatever forms they take, mammals' adaptations are among the aspects of their total personalities that people find most fascinating, perhaps because they are so immediately apparent. Therefore, this book emphasizes them heavily.

Skeleton and Teeth

All mammals have an internal structural support called an endoskeleton. It is the part of the body to which the muscles are attached. In addition to providing support, it protects certain internal organs. To do these jobs effectively, the skeleton must be extremely strong, and is in fact made of bone. It is also rather simple, partially as a result of many small bones having fused over the ages to make fewer relatively large structures such as the pelvic girdle. The skeleton is also fairly lightweight, which contributes to speed and the efficient use of energy.

Set into the jaws are the most durable structures of a mammal's entire body, the teeth. Because they are the last body parts to decompose, they are helpful in the study of fossils. In most cases they are also reliable clues to a mammal's diet, the teeth being well differentiated from each other and tailored to particular types of foods and life styles. For example, gnawing mammals such as beavers and many other rodents have highly developed incisors. The mouths of most carnivores are dominated by canines and carnassials, which are adapted for piercing and tearing flesh. And browsing and grazing mammals such as wild cattle and many other ungulates have well-developed premolars and molars (or cheek teeth, as they are referred to together). Molars themselves vary in the complexity of their grinding surfaces, those used for the fine pulverization of food usually being the most highly convoluted. All four types of teeth are present in the mouths of most mammals. But if a highly specialized type of tooth predominates, it usually determines diet.

Some insectivores cut only one set of teeth, but most mammals have two sets of teeth in the course of their lifetime. The first set, known as the milk teeth (or deciduous teeth), consists of incisors, canines, and premolars only. These are lost fairly early and are replaced by permanent dentition, which generally consists of incisors, canines, premolars, and molars.

Brain and Senses

The brains of mammals are large in comparison with those of other vertebrates. The largest part of the brain is the cerebrum, which is divided into two halves known as the cerebral hemispheres. Covering these hemispheres is gray matter technically known as the cerebral cortex. (The cerebral cortex is also referred to as the neopallium or neocortex—"new cortex"—since as it expanded over much of the original forebrain, it pushed the old cortex inward.) The neocortex has become intricately folded in many mammal species (and thus has increased in surface area), and in the placentals a complex of nerve fibers known as the corpus callosum has evolved to supply communication between the neocortex's two sides.

The neocortex is important primarily because it controls the learned behavior of mammals. More than any other part of the brain, it functions as the storage place for accumulated experience, and as the center for motor and intellectual activity. Stated differently, the neocortex is responsible for initiating and processing much mammalian behavior not considered strictly instinctive.

Most mammals' brains contain large olfactory bulbs and lobes, indicative of the importance of the sense of smell. In fact, in the vast majority of mammals smell is the dominant sense. This may be difficult for human beings to grasp, since smell, as a survival tool, is relatively unimportant to us. Yet sensitive noses no doubt make the world of other mammals seem very different from our own predominantly visual one.

Hearing is also well developed in most mammals, and has reached a high level of sophistication in many bats and cetaceans, which navigate and find their food through a system known as echolocation. This is a process by which certain species send out sounds that bounce off objects and back to the originator. Mammals are the only vertebrates with ear pinnae (external structures for funneling sound), and these are present in most aboveground species. In kangaroo rats and many other mammals that inhabit vast open spaces, the internal hearing mechanisms are enlarged to provide amplification.

The vibrissae, or whiskers, that most mammals bear near the nose or mouth are sensitive tactile organs. A well-developed sense of touch is especially important in the forest or the burrow (wherever light is dim), since most mammals have poor vision and cannot distinguish detail and color. This is in no way a handicap, however, since all mammal senses are appropriate to the life style of the individual species.

A given sensory stimulus, particularly a scent, may serve more than one purpose at a time. For example, the anal glands of some carnivores produce secretions that can serve functions as varied as territorial marking, self-defense, and sexual attraction. Often one individual will rub these secretions onto stones and branches, from which another individual "reads" them later on. Many ungulates have scent glands located in various parts of their feet or in front of their eyes. Secretions from these glands, as well as the urine that many animals deposit judiciously about their territories, are thought to serve a variety of purposes also.

Locomotion

Among mammals, quadrupedal (four-legged) locomotion is the norm. Bears, shrews, opossums, human beings, and many rodents place the entire foot on the ground each time they take a step, a type of movement described as plantigrade. An advantage of plantigrade movement to polar bears, for instance, is the sure-footedness it provides on ice. And among the much smaller species, where body weight is not such a concern, a flat-footed scurry is all that is needed.

Some mammals have evolved a method of moving known as digitigrade—they move on their digits (toes) all the time. This movement is most obvious, perhaps, in some of the larger carnivores such as the big cats. To them, speed is of the utmost importance in pursuing prey, and this type of movement allows for quick takeoffs and lightninglike sprints. But the large carnivores have the kinds of life styles that require them to use their extremities for other activities besides running—climbing trees, for example. Thus, they are equipped with supple paws and sometimes even with retractable claws.

The ultimate in streamlining for quadrupedal movement is a reduction in the number of toe digits, from five to three (as in the rhinoceros), two (as in the deer), or one (as in the antelopes). Among many hoofed mammals (ungulates), steady running is important. In fact, many ungulates spend practically their entire lives on their feet, and some migrate (at a relatively slow pace) over enormous distances. Except occasionally—as, for example, when they have to kick—these mammals do not need multifunctional extremities. Therefore a hoof, which is relatively stiff and unyielding, is appropriate. Moving on the hooves (which are actually the toenails) is described as unguligrade.

The order and speed with which quadrupeds place their limbs on the ground produces gaits as varied as a walk, a trot, a pace, and a gallop. The sequential placement of diagonally opposite limbs (right front, then left rear) typifies most quadrupedal walking. In the faster gaits stability is lessened as speed is increased—often only one foot is touching the ground at any given moment. And if the gait is quite fast, the animal's body may actually float in the air for nearly half the time. Although diagonal foot movement is the norm, camels, elephants, giraffes, and a few other mammals move both legs on one side of the body simultaneously.

Not all mammals are purists about their movements, of course; some combine a bipedal, or two-footed, style of locomotion with the more typical quadrupedal style. Certain primates walk upright for short distances while using their forelimbs to gather food. And kangaroos alternate moving on all fours with leaping on the hind legs only, the latter movement requiring them to use their tails for balance. Human beings are the only mammals that walk on only two feet for the great majority of their lives.

While quadrupedal terrestrial locomotion is the rule, many mammals never move this way at all, largely because they rarely or never touch the ground. Arboreal primates such as the gibbon, for example, are primarily brachiators. (Brachiation is a hand-over-hand kind of swinging through the branches.) Bats are the only mammals capable of true flight, and their specialized skeletons and flight membranes are among the marvels of the natural world.

Aquatic locomotion is not unique to aquatic species, but it is in them that the most beautiful streamlining occurs. Whales propel themselves at a speed of up to 15 knots by coordinating the movements of their horizontally fluked tails with those of their flippers. Their generally torpedo-shaped bodies, of course, help minimize water resistance. However, some aquatic mammals who spend

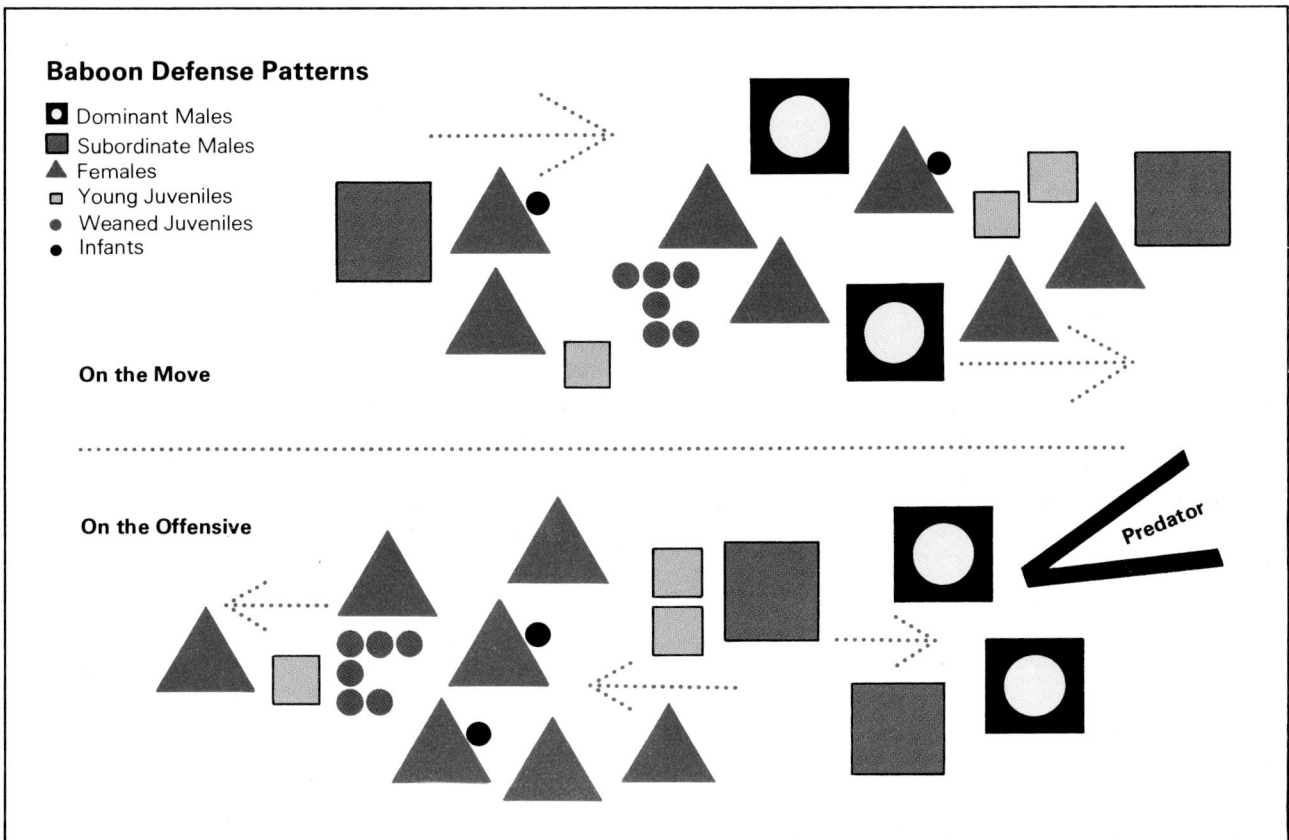

Baboon Defense Patterns

- Dominant Males
- Subordinate Males
- Females
- Young Juveniles
- Weaned Juveniles
- Infants

On the Move

On the Offensive

Predator

time out of the water—certain pinnipeds (seals and sea lions), for instance—move in an extremely labored manner on land. These mammals spend only a short time ashore each year, and their terrestrial movement is often as awkward as their swimming is lyrical.

Self-Defense

As a rule, no animal looks for trouble. Yet for most mammals there are daily occasions in which conflict arises in one form or another, and the only way out may be to run away from it. Most mammals, therefore, can move quickly enough to elude their major predators, at least for a while, if given sufficient warning time. Interestingly, however, predators must enter a critical range before they are considered threats to the potential prey. If they are outside this range, their presence triggers no flight response in the prey at all.

As an alternative to flight, the potential prey may respond to danger by remaining completely motionless. Usually the animal must be well camouflaged for this defense to work—rabbits and fawns, for example, use this

method of defense effectively. (Some mammals, in fact, change their coloration with the seasons and thereby remain inconspicuous year-round.) The opossum seems to carry the immobility tactic to its ultimate conclusion, that of feigning death. No one seems to know whether it is simply "playing possum" or actually in shock.

Specialized body parts play a large role in the self-defense of many species. Thick skin, for example, is a defense for many mammals, especially if it can also be rolled around softer body parts, as an armadillo's can. Tusks, horns, and sharp hooves and claws all offer valuable protection, as do the porcupine's quills. And sometimes an enemy can be so startled by a scream or some sort of warning grimace or posture that it is discouraged from even approaching a potential victim.

Social Life and Seasonal Patterns

Since at least minimal social contact is necessary for copulation and since every species must reproduce itself to survive, it is safe to say that all mammals engage in some degree of social activity. The extent and type of this activity varies greatly, however, from species to species. Many rodent males, for example, remain solitary except during breeding season. Then they seek out a mate, copulate, and depart—courtship as such is either minimal or nonexistent.

Some mammals such as the pocket gopher are considered colonial, since they construct their homes (burrows) in close proximity to one another. Yet the inhabitants of the different burrows really experience no social interaction, and in fact could be compared to human beings living in large urban apartment buildings. By contrast, other rodents such as prairie dogs create elaborate burrow systems, the members of which share food, alert each other to danger, and in many other ways have a very social life style.

Many of the more social, or gregarious, species live in hierarchical societies in which every member has its particular ''place'' somewhere along the dominance ladder. Although the more dominant places may be challenged from time to time, it is to the benefit of the entire group to keep the society running smoothly. Otherwise defense systems may break down, leaving the entire troupe vulnerable to devastation by predators.

The diagram on the preceding page illustrates how a baboon troupe organizes itself when moving to new feeding grounds and when on the offensive. The major responsibility for protecting the troupe lies with the dominant males. As the troupe moves, the most vulnerable positions—those at the front and the rear of the troupe—are taken by the subordinate males. The females and infants, at the center, are protected by the dominant males. In the attack formation, the dominant males position themselves at the front, leading the other males in an offensive against a predator, while the females and young stay at the rear, out of danger.

Although some mammals are monogamous throughout their lives, the vast majority are not. Much of the hoarding of females that goes on among harem animals is correlated with seasonal movements. Many of the ungulates, as well as seal and sea-lion bulls, are known for

maintaining a strong hold on definite territories, into which they herd as many females as they can attract and, hopefully, keep. The territories of some pinnipeds are established through combat among the bulls and are often defended through behavior that appears to be more ritual bluff than anything else—head shakes, stares, vocalizations, and other usually nonviolent displays.

True migration is a predictable movement back and forth between one (usually cold) region and another (usually warmer) area, with reproduction occurring in the latter. This definition does not encompass the mass emigrations of lemmings, since these are one-way journeys only. Nor does it apply to the permanent extension of an animal's range.

Although many bats and marine mammals migrate, the most spectacular examples of the phenomenon are probably seen in caribou and some other ungulates, which are restricted to land and must overcome enormous obstacles to reach their destinations. Caribou herds numbering up to a thousand individuals may travel some 375 miles (603 km) during their yearly journeys, generally south from the tundra along well-traveled routes into timbered areas in the fall, and then back to the tundra in spring, with some variations on this basic pattern. Some individuals starve along the way, and others drown or perish in other ways. Despite the dangers inherent in them, the movements continue year after year.

If superior feeding grounds, warmer weather, and congenial conditions are the motivating factors for migration, it is much less clear what the actual mechanisms controlling the phenomenon are. In birds, environmental conditions such as temperature and length of daylight are known factors, particularly since they influence the pituitary, which is closely linked to gonadal development and metabolism. (These determine physical preparedness for migration.) Among mammals, however, food shortages are the only factors that are definitely known to trigger migration.

Mammals that do not migrate may be adapted to withstand stressful environments by hibernating, estivating, or going into some sort of temporary torpor. Like true migration, true hibernation is a relatively rare phenomenon, and is not much better understood. Among mammals, it occurs only in the bats, insectivores, and rodents. Strictly speaking, hibernation refers to that body state in which heartbeat and temperature drop dramatically and breathing becomes irregular. Usually a hibernating mammal cannot be roused unless it is shaken violently, and even then it may not respond, since its body temperature may be as low as 37 to 39° F (3–4° C).

The primary advantage of hibernation is clear—it allows for decreased energy expenditure during very cold seasons, when food supplies are generally inadequate. But just what triggers it is certainly not clear. It may be related to the length of daylight, to fat levels built up in the animal's body over late summer and autumn, or to any number of other factors acting in combination.

The Fossil Record
Fossil evidence shows that enormous changes have taken place in the animal life of this planet. What started out as small, simple marine invertebrates more than 500 million years ago gradually evolved into vertebrates. Although fishes were the first real vertebrates, the development of a backbone made life on land possible, and the amphibians gradually evolved from the fishes. Though still tied to the water for breeding, the amphibians began to come ashore for their other activities. From them the reptiles gradually evolved, and from some of the reptiles the mammals eventually sprang.

The Tertiary period, which began some 60 to 70 million years ago, is generally recognized as the Age of Mammals, though the line between the most advanced reptiles and the earliest mammals was undoubtedly not very clear. The close of the Mesozoic era, immediately preceding the Tertiary period, was marked by pronounced changes in climate and other forces with which warm-blooded animals were better equipped to cope than were the dinosaurs and other previously dominant reptiles. As a result, the latter died out, and the mammals filled

Table of Geological Time
Began (mill. yrs.) — *Typical Animals*

Era	Period	Epoch	Began (mill. yrs.)	Typical Animals
Mesozoic Era	Triassic Period		230	Mammallike reptiles
	Jurassic Period		180	Dinosaurs, pterodactyls
	Cretaceous Period		140	Dinosaurs, marsupials, placentals
Cenozoic Era	Tertiary Period	Paleocene Epoch	65	Multituberculates, pantodonts
		Eocene Epoch	55	Archaic ungulates and carnivores
		Oligocene Epoch	36	Early apes, primitive mastodons
		Miocene Epoch	25	Bear-dogs, dryopithecine apes
		Pliocene Epoch	12	*Hipparion* (3-toed horse)
	Quaternary Period	Pleistocene Epoch	3	Primitive men, mammoth
		Holocene Epoch	01	Modern man, domestic animals

the gaps that they had vacated. By the Eocene epoch most of the present mammalian orders had evolved.

The Ecology of Mammals
Rather than viewing and labeling individual organisms as if they were isolated and existing in a vacuum, many scientists now examine plants and animals in the context of their total environment. Thus they are engaged in the study of ecology, the science of living things in relation with their surroundings and each other. The recent concern with endangered species and worldwide environmental quality has made ecological thinkers out of many nonscientists as well. As a result, ecological terminology is familiar to many people by now, and is mentioned here only briefly.

Food Chains

The physical, or nonliving, environment interacts with the living environment to form what is known as an ecosystem. This is any self-contained, self-perpetuating natural system, regardless of size. The physical components of an ecosystem are air, water, sunlight, and minerals. The living components (plants and animals) are the food producers, the food consumers, and the decomposers—those organisms that break down organic matter and convert it to inorganic nutrients that can be returned to the environment.

Through the miracle of photosynthesis green plants convert the energy in sunlight into food, first into a simple sugar known as glucose. Glucose becomes dissolved in sap and moves through the entire plant, nourishing the plant as it travels, in much the same way that the nutrients dissolved in blood nourish animals. Some glucose is changed into starch, some combines with other materials to make fats, proteins, and other types of foods. Regardless of its form, the food made by green plants is the ultimate source of food for the entire animal world.

Animals that eat this food directly are known as first-order consumers. Many rodents, dozens of ungulates, and numerous other species fall into this category. The word *herbivore* means plant-eater, and is another way of referring to first-order consumers.

Those animals that eat the animals that eat the plants are referred to as second-order consumers, or as *carnivores* (meat-eaters). Some carnivores are specialized to eat only one type of animal food—insects, for example. In that case they are predictably called insectivores. (Similarly, some herbivores eat primarily fruit, and are referred to as frugivores.) Those animals on any level with a very generalized diet—that is, a diet that includes both plants and animals—are referred to as *omnivores.*

This type of what-eats-what relationship can progress through several stages until the final consumer emerges as that predator with no natural enemy that preys upon it regularly. (Thanks to modern weaponry, man is most often the final consumer today.) Ecologists

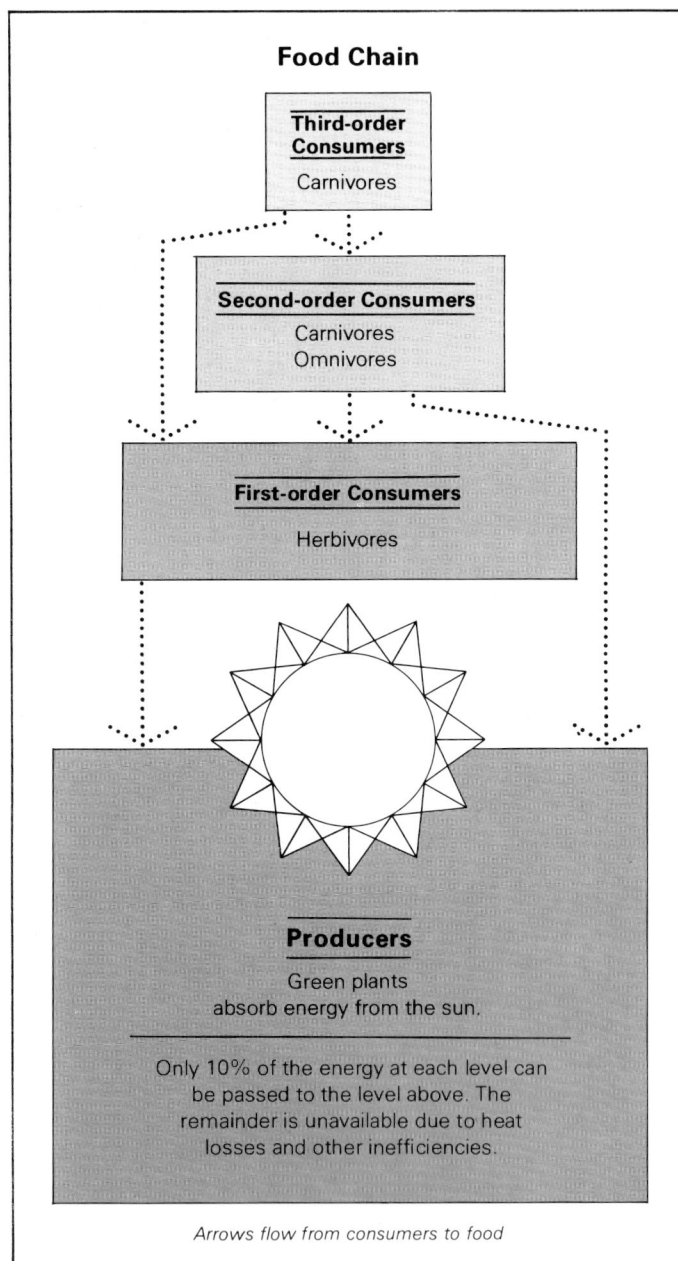

Food Chain

Third-order Consumers
Carnivores

Second-order Consumers
Carnivores Omnivores

First-order Consumers
Herbivores

Producers

Green plants absorb energy from the sun.

Only 10% of the energy at each level can be passed to the level above. The remainder is unavailable due to heat losses and other inefficiencies.

Arrows flow from consumers to food

use the metaphor of a food chain to represent these interactions, and the final consumer is described as being at the top of the food chain.

The total number of individuals at each level in a food chain (or at each trophic level, as it is sometimes called) is by no means a matter of chance. As a rule plants, at the bottom of the chain, are abundant, and prey species are more numerous than their predators—their populations must be large enough to spare some individuals without the entire species suffering. The number of offspring a given species bears is determined, therefore, at least in part by that species' position on a food chain, the smaller and more vulnerable species usually bearing the most young.

Of particular interest to people concerned with world food supplies is what happens to energy as it is passed (in the form of food) from one trophic level to the level above it. At each level, animals disperse some energy through their activities, making that energy unavailable to consumers on the next trophic level. (In fact, 90 percent of the energy available at each trophic level is dispersed.) By the time energy is transferred through several levels, then, the greater part of it has been dispersed. The implications here are startling, at least for human populations. Since less energy is available at higher trophic levels, the inefficiency of eating cattle (a first-order consumer) as opposed to grain (a producer) becomes obvious.

Adaptive Radiation

This book presents mammals in such a way as to show how their adaptations equip them for a certain environment. Within the environment, the habitat is the particular set of physical conditions in which a species or a community of animals lives. A woodland is a habitat, for example. What a species does within its habitat (in other words, the particular role it performs) is referred to as that species' ecological niche. Within the woodland, a gray squirrel's niche is that of a diurnal (daytime) arboreal (tree-dwelling) herbivore (plant-eater). By occupying the niche it does, the squirrel provides benefits to the community—for example, it disperses acorns, from which oak trees grow.

The squirrel and other mammals have come to occupy their particular niches through a very long process known as adaptive radiation. Basically this is the process that living things undergo as they evolve over generations from an ancestral stock, generalized in form and behavior, into creatures much more specialized in particular directions. For example, fossil evidence suggests that the mammals' ancestral stock was an animal very much like a contemporary insectivore, with unspecialized teeth, five claws on each of four limbs, and few if any limb specializations. From this primitive stock some mammals began specializing in particular ways. Some became gnawers, the forerunners of the rodent order. Others became runners—their digits gradually became simplified, eventually evolving into hooves. (These mammals gave rise to our modern ungulates.) Fliers, swimmers, and many other specialists also evolved from the basic insectivorelike stock, and each type of specialty is roughly represented in the nineteen extant mammal orders.

The Mammal Orders and Continental Change

Today there are nineteen mammal orders. The monotremes (comprising the platypuses and the echidnas) are considered the most primitive order (that is, the one that evolved earliest) because they lay reptilelike eggs—soft, pliant, and leathery—and because a single chamber known as the cloaca is used for both reproduction and elimination. In other mammals there are separate chambers for these different functions. Because of this type of evidence, the monotremes are thought to have diverged early on from the mainstream of mammal evolution, and are not considered the forebears of the more advanced mammal species.

Marsupials, from which present-day mammals *are* thought to be descended, reproduce in a different manner. The females carry their embryos for a very short while before they bear them live. This means that relative to most mammals' offspring, marsupial newborns are only partially developed at birth, and continue maturing for quite some time thereafter inside their mother's pouch.

Mammal Classification

Order names in **boldface** Family names in *italics*

Monotremata: Monotremes or
Egg-laying Mammals
Tachyglossidae: Spiny Anteaters
or Echidnas
Ornithorhynchidae:
Duck-billed Platypuses

Marsupialia: Marsupials or
Pouched Mammals
Didelphidae: American Opossums
Dasyuridae: Pouched or Marsupial
"Mice" and "Rats,"
Native "Cats," Tasmanian Devils,
Tasmanian "Wolves"
Myremecobiidae: Numbats
Notoryctidae: Marsupial or
Pouched "Moles"
Peramelidae: Bandicoots
Caenolestidae: Rat Opossums
Phalangeridae: Phalangers;
Cuscuses; Possums; Koalas
Phascolomydae: Wombats
Macropodidae:
Kangaroos; Wallabies

Insectivora: Insect-eating Mammals
Solenodontidae: Solenodons
Tenrecidae: Tenrecs;
Madagascar "Hedgehogs"
Potamogalidae: Water Shrews;
Otter Shrews
Chrysochloridae: Golden Moles
Erinaceidae: Moon Rats;
Gymnures; Hedgehogs
Macroscelididae: Elephant Shrews
Soricidae: Shrews
Talpidae: Shrew-moles;
Desmans; Moles

Nesophontidae: Extinct West
Indian Shrews

Dermoptera: *Cynocephalidae:*
Gliding or "Flying" Lemurs

Chiroptera: Bats
Pteropodidae: Old World Fruit Bats
Rhinopomatidae: Mouse-tailed Bats
Emballonuridae: Sheath-tailed Bats;
Sac-winged Bats;
Tomb Bats
Noctilionidae: Fisherman Bats
Nycteridae: Slit-faced Bats;
Hollow-faced Bats
Megadermatidae: False Vampire
Bats; Yellow-winged Bats
Rhinolophidae: Horseshoe Bats
Hipposideridae: Old World
Leaf-nosed Bats
Mormoopidae: Mustached Bats
Phyllostomatidàe: American
Leaf-nosed Bats
Desmodontidae: Vampire Bats
Natalidae: Funnel-eared Bats
Furipteridae: Smoky Bats;
Thumbless Bats
Thyropteridae: Disk-winged Bats
Myzopodidae: Sucker-footed Bats
Vespertilionidae: Vespertilionid Bats
Mystacinidae: New Zealand
Short-tailed Bats
Molossidae: Free-tailed Bats;
Mastiff Bats;
Naked Bats

Primates: Primates
Tupaiidae: Tree Shrews

Lemuridae: Lemurs
Indridae: Avahis; Simpoonas; Indris
Daubentoniidae: Aye-ayes
Lorisidae: Lorises; Pottos; Galagos
Tarsiidae: Tarsiers
Cebidae: New World Monkeys
Callithricidae: Marmosets; Tamarins
Cercopithecidae: Old World
Monkeys and Baboons
Pongidae: Gibbons; Orang-utans;
Chimpanzees; Gorillas
Hominidae: People

Edentata: Sloths; Anteaters;
Armadillos
Megalonychidae: Extinct Ground
Sloths
Mylodontidae: Extinct Ground Sloths
Myrmecophagidae: Anteaters
Bradypodidae: Tree Sloths
Dasypodidae: Armadillos

Pholidota: *Manidae:* Pangolins

Lagomorpha: Pikas; Rabbits; Hares
Ochotonidae: Pikas
Leporidae: Rabbits; Hares

Rodentia: Rodents
Aplodontiidae: Sewellels;
Mountain Beavers
Sciuridae: Squirrels;
Woodchucks; Prairie Dogs;
Chipmunks
Geomyidae: Pocket Gophers
Heteromyidae: Pocket Mice;
Kangaroo Mice; Kangaroo Rats
Castoridae: Beavers

Adapted from Ernest P. Walker *et al., Mammals of the World,*
3rd ed. Revised for Third Edition by John L. Paradiso. © 1964, 1968, 1975
by The Johns Hopkins University Press, Baltimore.

Anomaluridae: Scaly-tailed Squirrels
Pedetidae: Springhares
Cricetidae: New World
    Rats and Mice;
    Hamsters; Voles;
    Lemmings; Gerbils
Spalacidae: Mole-rats
Rhizomyidae: African Mole-rats;
    Bamboo Rats
Muridae: Old World Rats and Mice
Myoxidae: Dormice
Platacanthomyidae:
    Spiny Dormice;
    Chinese Pygmy Dormice
Seleviniidae: Desert Dormice
Zapodidae: Birch Mice;
    Jumping Mice
Dipodidae: Jerboas
Hystricidae: Old World Porcupines
Erethizontidae: New
    World Porcupines
Caviidae: Guinea Pigs; Cavies
Hydrochaeridae: Capybaras
Dinomyidae: Pacaranas; False Pacas
Heptaxodontidae: Seven Extinct
    Genera of the Greater and
    Lesser Antilles
Dasyproctidae: Pacas; Agoutis
Chinchillidae: Viscachas; Chinchillas
Capromyidae: Hutias; Coypus
    or Nutrias
Octodontidae: Chozchoris; Coruros;
    Rock Rats; Viscacha Rats
Ctenomyidae: Tucu-tucos
Abrocomidae: Chinchilla Rats
Echimyidae: Spiny Rats
Thryonomyidae: Cane Rats
Petromyidae: Rock Mice or

Rock Rats
Bathyergidae: African Mole
    Rats; Blesmols
Ctenodactylidae: Gundis;
    Speke's Pectinators

Cetacea: Whales; Dolphins;
    Porpoises
Platanistidae: Freshwater or
    River Dolphins
Ziphiidae: Beaked Whales;
    Bottle-nosed Whales
Physeteridae: Sperm Whales
Monodontidae: Belugas; Narwhals
Delphinidae: Dolphins; Porpoises;
    Killer Whales; Pilot Whales
Eschrichtiidae: Gray Whales
Balaenopteridae: Rorquals;
    Humpback Whales;
    Blue Whales
Balaenidae: Right Whales

Carnivora: Carnivores
Canidae: Dogs; Dingoes; Jackals;
    Wolves; Coyotes; Foxes
Ursidae: Bears
Procyonidae: Cacomistles;
    Raccoons; Coatimundis;
    Kinkajous; Olingos; Pandas
Mustelidae: Weasels; Polecats;
    Ferrets; Minks; Martens;
    Tayras; Grisons; Wolverines;
    Badgers; Skunks; Otters
Viverridae: Civets; Mongooses;
    Suricates; Fossas
Hyaenidae: Hyaenas; Aardwolves
Felidae: Cats; Lynxes; Bobcats;
    Pumas; Jaguars; Lions; Tigers

Pinnipedia: Seals; Sea Lions;
    Walruses
Otariidae: Sea Lions; Fur Seals
Odobenidae: Walruses
Phocidae: True, Earless, or Hair Seals

Tubulidentata:
    Orycteropodidae: Aardvarks

Proboscidea: *Elephantidae:*
    Elephants

Hyracoidea: *Procaviidae:* Hyraxes

Sirenia: Dugongs; Manatees
Dugongidae: Dugongs; Sea Cows
Trichechidae: Manatees

Perissodactyla: Odd-toed Ungulates
Equidae: Horses; Zebras; Quaggas;
    Asses; Onagers; Kiangs
Tapiridae: Tapirs
Rhinocerotidae: Rhinoceroses

Artiodactyla: Even-toed Ungulates
Suidae: Hogs; Pigs
Tayassuidae: Peccaries
Hippopotamidae: Hippopotamuses
Camelidae: Camels; Guanacos;
    Llamas; Alpacas; Vicuñas
Tragulidae: Chevrotains; Mouse Deer
Cervidae: Deer; Muntjacs; Moose;
    Elk; Caribou
Giraffidae: Giraffes; Okapis
Antilocapridae: Pronghorn Antelopes
Bovidae: Cattle; Buffalo; Bison;
    Oxen; Duikers; Antelopes;
    Gazelles; Goats; Sheep

Usually they attach themselves to a nipple as soon as they get into the pouch, and are thus continually provided with food.

In all mammals other than monotremes and marsupials, embryos are carried until completely formed, in a placenta attached to the mother's uterus. The vast majority of mammals are thus known as placentals. What "completely formed" means, of course, varies enormously, depending on the environmental demands placed on each species—mammals that excavate dens, for instance, can afford to bear more dependent young than ungulates that are constantly on the move. In general, mammal newborns are developed to the extent that they can continue the maturation process without benefit of egg or pouch, and thus become a part of mammal "society" while still very young.

Primates are usually considered the most advanced of all placentals, and therefore of all mammals. Among the several characteristics that have earned them this distinction are the strong binocular vision, opposable thumb, and highly developed brain that are evident in the more advanced species. Human beings are primates, of course, primates that differ from all the others by virtue of upright bipedal locomotion and, much more importantly, a particular type of mental activity. Only human beings are believed capable of conceptual thought, that is, of the ability to think in abstract terms instead of simply responding to stimuli. And only human beings are thought to be aware of their own consciousness. Such faculties are responsible for the entirety of human culture.

The mammal orders evolved at a time when strong geological forces were at work on the face of the earth, and these forces influenced where particular orders arose. Past movements of the continents themselves were among the strongest of these forces, and dramatic changes in climate resulted from these movements. For example, judging from glacial flora and other types of evidence, it seems clear that what are now the earth's more northern continents—North America, Europe, and Asia—plus Greenland, were at one time joined to form one large land mass that biogeographers call Laurasia. The other continents—Africa, South America, Australia, and Antarctica—plus India, formed another land mass, now known as Gondwanaland. The force of convection currents bringing hot matter from the earth's interior to its cooler surface gradually fragmented these two supercontinents, and the fragments became today's continents. As a result of fragmentation, land masses assumed new positions relative to the equator and the poles, and the oceans filled the gaps between the newly separated continents. These changes led to greater variations in climate than there had been before.

The beginning of the breakup of Laurasia and Gondwanaland preceded by many, many years the beginning of the radiation of mammals from primitive forms into various orders. Thus, except possibly where there were still remnants of land connections, such as the land thought to have connected northern Asia and North America across the Bering Strait, each continent was already isolated from the others when the radiation began. Consequently each continent developed its own distinctive fauna. The fauna of North America and Asia, however, are remarkably similar, quite possibly due to the Bering connection just mentioned, which may have served as a bridge across which mammals could pass.

South America became isolated from North America fairly early in the Tertiary period, the connection that exists today not appearing until much later. Thus it developed a distinctive fauna made up of the edentates (anteaters, sloths, and armadillos) plus, possibly, the opossum and other marsupials. The origin of marsupials, in fact, is controversial. Some authorities believe that marsupials were fairly widespread before the better-adapted placentals overtook them everywhere except on the totally isolated island of Australia. Other authorities claim that the marsupials actually originated on Australia, where they have radiated out to fill the same niches that the placentals occupy elsewhere.

Africa's native fauna includes the elephants, hyraxes, aardvarks, and several extinct orders. The exact origin of

Zoogeographical Realms

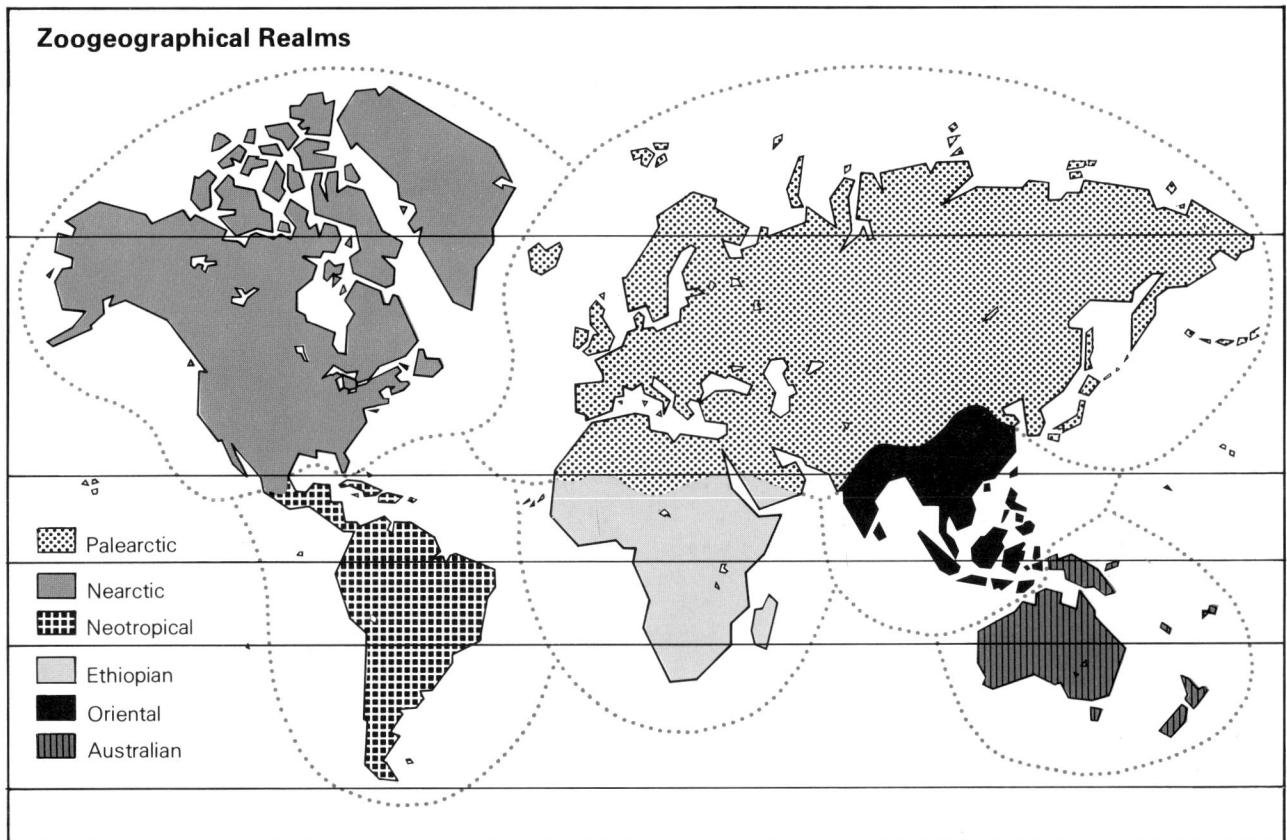

Palearctic

Nearctic

Neotropical

Ethiopian

Oriental

Australian

many of the primates and other mammal orders has been obscured by the many land connections that are thought to have arisen and disappeared between the various continents, as well as by the fact that some mammals (bats, for instance) can travel great distances.

Dividing the world into what are known as zoogeographical realms, as shown on the map here, provides a convenient way of referring to each of the distinctive faunas that are the basis for the nineteen mammal orders. As one would expect, the realms—the Neotropical, the Nearctic, the Palearctic, the Ethiopian, the Oriental, and the Australian—coincide roughly with the continents. Although the mammals of Madagascar are quite distinct, that island is often treated as part of the Ethiopian realm.

The Species and Competition

A species is a group of living things that have become so specialized in the same ways that they breed only among themselves. A species cannot permanently tolerate any other species occupying its ecological niche at the same time and in the same place. Over the ages, species have evolved in various ways that ensure that they alone occupy a niche without actually having to fight other species over it. That is, competition is reduced by means that do not necessarily involve direct confrontation.

A common way of reducing competition for niches is for some populations to be adapted to feed by day while others are adapted to feed by night; although the different species may prey on the same populations of smaller animals, they lessen competition with each other by doing it on different schedules. Another way is for different species to have differently adapted body parts—legs or necks, for example, may be different lengths. The giraffe and several other African ungulates may all feed on the same trees, but they do it from different heights and so avoid competition. Because competition for niches is reduced wherever possible, it is difficult to see it at work. Nonetheless it is a continuing process that operates to maintain species as they are or to change them.

The same niche is often filled by one species in one

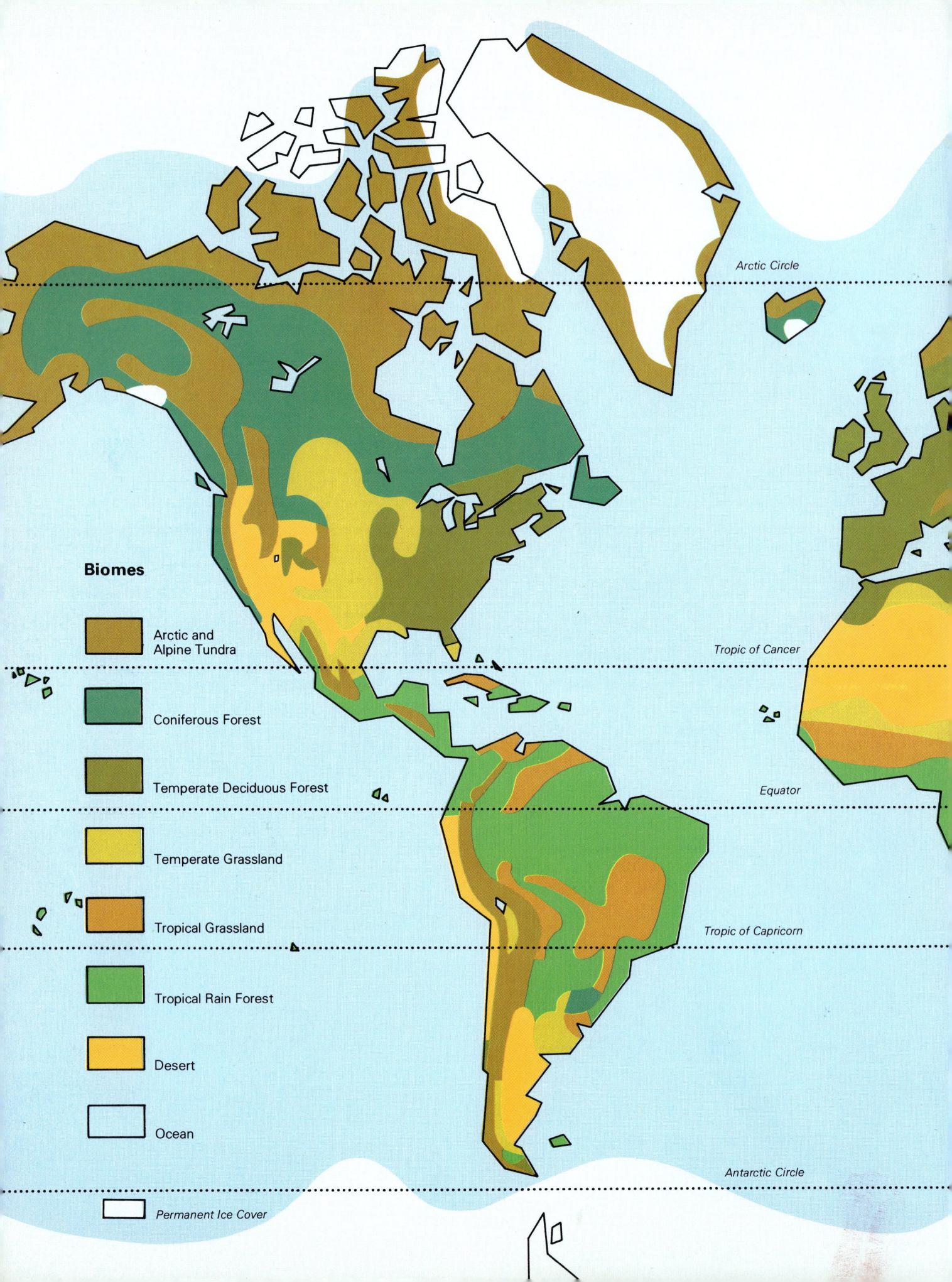

Biomes

- Arctic and Alpine Tundra
- Coniferous Forest
- Temperate Deciduous Forest
- Temperate Grassland
- Tropical Grassland
- Tropical Rain Forest
- Desert
- Ocean

Permanent Ice Cover

Arctic Circle

Tropic of Cancer

Equator

Tropic of Capricorn

Antarctic Circle

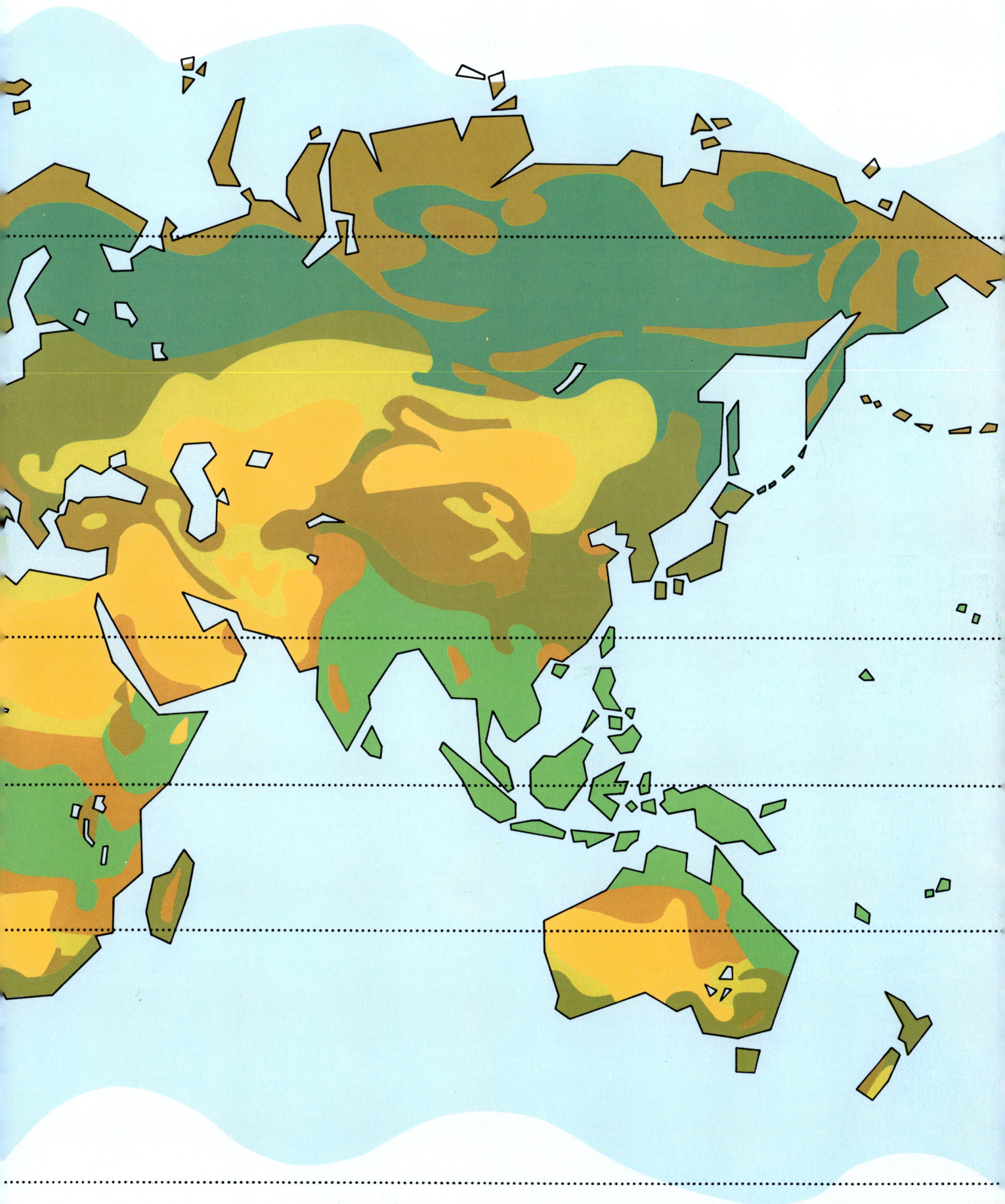

locale and by a different species in another locale. When this happens, the species are referred to as each other's ecological equivalents, or as products of a process known as convergent evolution. For example, the anteaters of South America, the spiny anteaters of Australia, the pangolins of Africa and Asia, and the aardvarks of Asia—all unrelated species—are each other's ecological equivalents. Similar environmental pressures on their respective continents resulted in the same adaptations eventually converging in the four animals. All are relatively large terrestrial insect-eaters with a narrow, sticky tongue, an elongated snout, characteristic posture, size, and speed, and many other features in common.

Biomes

Latitude, altitude, and position relative to oceans all determine climate. And climate determines the dominant vegetation of a region, which is of primary importance to the mammals and other animals found there. The earth's major climatic zones are called biomes, and each has its characteristic assembly of animals and plants. (Most biomes are named for their dominant plant growth, in fact.) Although biogeographers are not in total agreement as to the number of biomes, most recognize between seven and ten terrestrial ones, plus the ocean. This book describes the following eight: the tropical rain forest, the temperate deciduous forest, the coniferous forest, the tropical grasslands, the temperate grasslands, the desert, the tundra, and the ocean.

How This Book Is Organized

What constitutes the real heart of this book are the fascinating and practically endless ways in which mammals are adapted to their particular surroundings. Since the biome is the largest and most general unit used to describe those surroundings, it provides an easy way of organizing the world's mammals into groups of manageable size—desert mammals, tundra mammals, and so forth. And it has an advantage over a strictly geographical organization (particularly since there is so much disagreement

over which mammals are native to the various continents) in that it allows for many types of comparisons and contrasts among totally unrelated species now living in different parts of the world. Mammals find their food, seek mates, rest, and carry out their myriad other activities in the particular ways they do primarily because they live in a certain biome (and in a certain habitat within that biome). Therefore describing mammals in the context of what determines their life styles—that is, the biomes—provides a flexible means of organization with a strong emphasis on interaction with the environment.

The book is divided into eight sections, one section for each biome. Each section has its own introduction, which sets out first the physical and vegetational characteristics of the biome, and second, the general ways in which typical mammals are adapted to these conditions. Elaborations on the adaptations appear in the pieces on individual mammal species, which make up the real body of the text.

Within each biome the mammals are organized according to geography, with the largest and most spectacular regions appearing first. For example, within the tropical grasslands section, Africa is presented first, since it contains the finest example of this biome in the world. Within each geographical section the mammals are generally presented in phylogenetic order, that is, in the order in which they are thought to have evolved. This means that the primitive species are presented before the more advanced ones, hopefully with the result that the reader can trace some of the important developments in the evolution of the mammal class as a whole. The particular species included in this book were chosen on the basis of how typical of their biome they are, and of how well they balance the other species to form a representative fauna, or characteristic group of animals.

Although some mammal species inhabit more than one biome, they are usually presented here in the context of the one that is most typical of them. The order in which the biomes are presented is somewhat arbitrary.

Terms such as "chiropter" (bat), for example, are

usually anglicized forms of the names of the orders (Chiroptera), and can be easily understood if the reader refers to the classification table included in this introduction.

Endangered Species

"Endangered," not "handsome" or "intelligent," is all too often the adjective that precedes the word "species" today.* Its exact meaning is difficult to understand, since each species has its own particular requirements for survival. Generally, the word is applied to any species whose total population has been reduced to a number of individuals very near the minimum number required for that species to sustain itself. In other words, if the population were to become much smaller, the species would probably become extinct. The minimum number itself varies from species to species, being higher in the more social mammals and lower in the more solitary ones. And other factors are also often considered in determining whether a species is endangered—condition of habitat, frequency of disease, and protective measures in effect, for example.

The Survival Service Commission of the International Union for Conservation of Nature and Natural Resources collects information on endangered plant and animal species throughout the world. It also publishes lists of those species and tries to encourage action to aid in their preservation. Its publication, known as the *Red Data Book*, is the official check list for endangered wildlife. A looseleaf collection that is periodically updated, it contains information concerning the nature of the threats to a species' continued survival, that species' habitat, distribution, and population (if known), proposed conservation measures and measures already in effect, and exact status—endangered, vulnerable, or rare.

Endangered wildlife is a subject that cannot be viewed dispassionately—at least 120 mammal species are currently threatened with extinction, nearly 80 percent of them most probably because of human activity and not simply because of natural phenomena. Governments are responsible for drastic losses suffered by the wolf, for example, since even after it was proved an ex-

tremely peaceable creature with strong family bonds, it continued to be persecuted, with official sanction, on a widespread scale. Many individuals such as whalers make their livings directly from animals' deaths, and often trophy hunters, furriers, and a host of other people play a significant part in threatening some portion of the world's wildlife.

The only difference between these people and the rest of us is that we are removed from the actual poisoning, trapping, and slaughter. Yet though we may not realize it, we are often just as guilty. When we eat tuna, for example, we may also be threatening dolphins, since more than 100,000 dolphins a year are killed as a result of current tuna-fishing practices. And habitat destruction, for which many people are responsible, is perhaps the worst offender of all.

Because the damage we do is so often done in innocence, it must be brought to light, and communal pleas to examine the consequences of many of our actions, often pleas with particular species in mind, are now being heard worldwide. Even where threats to a species' continued survival are extremely subtle, as in the case of habitat destruction, there now exist groups of concerned individuals working hard to stem the tide of extinction.

The reason that the issue of endangered species is such a grave one has to do with much more than esthetics and discouraging the kind of mindlessness required to destroy quite willingly the products of millions of years of evolution. Those factors are certainly valid, but a more immediate reason for stemming the tide of endangered species is to ensure the future of the entire living world. What is good for the wild creatures of this planet is also necessary for human beings—physical space and enough green plants for the production (through photosynthesis) of food to eat and oxygen to breathe. If we allow these basic elements to be replaced by endless highways and housing developments for our burgeoning human numbers—that is, if we continue to disrupt our *own* natural habitat—we can eventually enter *Homo sapiens* in the *Red Data Book* and simply close the cover.

---

* Endangered species are noted in the text with an asterisk (*).

1

The Tropical Rain Forest

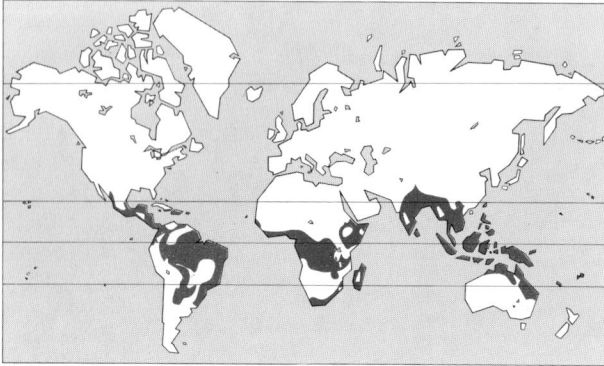

Tropical rain forests lie between the tropics of Cancer and Capricorn on five continents—from the southernmost part of North America into northwestern South America and the entire Amazon basin; in the western half of equatorial Africa and in eastern Madagascar; and in most of the Indo-Malayan archipelago into northeastern Australia. The biome is characterized by a year-round temperature of about 81° F (27° C) and by extremely high humidity and rainfall, over 79 inches (200 cm) annually and no less than 5 inches (12 cm) any month. This marked uniformity of climate is responsible for the rain forest's year-round growing season. Plants germinate, flower, bear fruit, and die throughout the year.

The calcium and other nutrients present in rain-forest soils undergo an extremely rapid turnover, for several reasons. The heavy rainfall itself has a leaching effect on soil, and the structure of the plant leaves, which are evergreen, contributes significantly to the leaching process. Typically, rain-forest trees have oblong leaves with "drip tips," tapered ends designed to shed water efficiently. In addition, decay proceeds so rapidly in this moist biome that humus is almost completely lacking. And the numerous termites and ants that populate the rain-forest floor have extremely efficient digestive systems. Their digestion is so rapid that they consume much of the ground litter before it can become part of the soil.

All of these climatic and physical forces working to-gether over time have resulted in the richest and most complex plant community on earth. The tropical rain forest has by far the most plant species of any biome, around 100,000 of them. Many of them resemble each other to a startling extent, yet no one of them is found in large concentrations, perhaps because the slight differences in soil composition from one locale to another demand equally slight but significant differences in the plants that those soils support. Over a given area of rain forest, therefore, many closely related yet distinctly dissimilar plant species usually grow side by side, and groves of like individuals occur infrequently.

The rain forest is so "arranged" structurally that it is possible to discern several layers of plants in it, at least two or three woody layers—the trees and shrubs—plus the few herbaceous plants that make up the undergrowth. The uppermost layer is generally known as the emergent layer, with trees that measure between 148 and 180 feet (45–55 m) in height and stand high above the rest of the forest. Most emergent trees, and the shorter ones as well, have slender trunks that branch into crowns very high and are often supported by buttress-like formations. Their bark is generally smooth, thin, and light in color.

This uppermost layer and the layer just below it (which is sometimes called the canopy layer) form an unbroken mass of greenery in which some mammals spend their entire lives. In fact, these two layers together constitute the real stage on which most rain-forest activity takes place. This is so for a very obvious reason—light. Much of the light available to the two upper levels is blocked before it reaches the lower ones. And the lowest layer of mature trees must compete for light with vast numbers of saplings, not to mention with the palms, ferns, grasses, and other types of plants that make up the undergrowth. Hence many tropical fruits and flowers are found at considerable heights, with the result that animals must often forage high in the branches to satisfy their need for vegetable foods.

In addition to the many trees and other freestanding

Preceding pages: Young orang-utan, Sumatra

plants of this biome, there are scores of plants that entwine themselves in fantastic patterns over and among the branches of some host, and are thus considered dependent species. They, too, are seeking access to the light. Like the trees, these plants are generally woody. Some of them, such as the climbers rooted in the soil, are known as lianas. Others go by the general name of epiphyte, and take their nutrients not from the soil, but from rain water, air, and whatever debris accumulates near their roots.

Many rain-forest plants are transitional between self-supporting trees and vines, and also between climbers and epiphytes. For example, the Malayan orchid *Dipodium pictum* begins its life with ground roots, then gradually becomes an epiphyte. Stranglers such as the strangler fig progress in the opposite direction. That is, they start out as epiphytes, and as they mature, send roots down to the ground and into the soil. Often they kill the host tree that originally supported them, and become independent plants in the process.

The agents of decay are the saprophytes and parasites such as the numerous fungi that are scattered over the forest floor. They are directly responsible for the fast decomposition of fallen trees and limbs and of the sparse ground litter. Since they do not manufacture their own food, they do not need to compete for available light. In fact, these relatively small plants are practically the only nonentrants in the race upward toward sunlight—except where fallen trees have opened the canopy, the ground is almost devoid of greenery.

The tropical forest is important to the life of this planet in a very special way. Its luxuriant vegetation is indicative of its position, relative to the other biomes, in the production of the earth's supply of oxygen. But the rain forest's natural vegetation is being destroyed at an alarming rate. In many areas it is being replaced by plant communities capable of producing food for human beings, but not capable of producing oxygen comparable to the original level of production. If unchecked, such reductions of photosynthesis—and the oxygen depletion that results

from such reductions—can condemn much, if not all, of the planet's life to oblivion.

The high sunny branches of tropical trees may be an embarrassment of fruiting or flowering riches at certain times. Yet each species of tree is on its own production schedule and, as mentioned earlier, is typically surrounded by other trees unlike itself. This means that although vegetable food is available year-round to inhabitants of the canopy layer, it is generally scattered, and those mammals that feed on it must be prepared to take advantage of it when and where it ripens. To enable them to do this, their body size and shape must allow them to travel easily, and they must have senses, appendages, and societies that help them locate the bounty and then partake of it. This is no real hardship, especially not in comparison with the living conditions that tundra and desert mammals face. Yet certain adaptations are necessary for survival even in the midst of plenty, and the many rain-forest mammals are specialized in a multitude of ways.

Primates are the quintessential rain-forest mammals. They feed almost exclusively in the trees; some of them are equipped to leap, while others swing from branch to branch on their very strong forelimbs, a means of locomotion known as brachiation. Still others, such as the lemurs, run along the top surfaces of the branches on all fours. Some New World primates (which are much more arboreal, as a group, than the Old World primates) have prehensile, or grasping, tails, which act as a fifth limb when the animals climb. These are usually present only in the slower climbers, as a kind of insurance against falls. Arboreal primates also have opposable thumbs and big toes, an adaptation that allows them to wrap their hands and feet around branches. And the gripping power of the more highly evolved primates is further enhanced by flattened digits typically accompanied by nails.

The higher primates also possess binocular vision, important for depth perception in animals that leap from tree to tree, and they have color vision. A keen sense of hearing is also essential in the treetops, since dense

foliage makes seeing over long distances impossible.

Primates and other vegetarians of the rain forest's upper layers have evolved social structures and systems flexible enough to permit them to follow the changing food supplies. Some of them are more or less permanently nomadic, wandering from one fruiting tree to the next. Other social species have evolved sedentary groupings small and undemanding enough to have a minimal impact on food supplies. Many species in the latter type of group maintain territories, but they may also become nomads if conditions demand it.

In addition to the primates, many other types of mammals populate the rain forest's upper layers. In Central and South America, for example, sloths eat, sleep, and move upside-down for most of their lives, being adapted to hang by their strong, hooklike claws. The tree hyraxes of Africa are equipped with special adhesive padding on the soles of their feet, which allows them to scale trees and build nesting holes in the higher parts of the trunks. Even various insectivorous mammals—some of the anteaters and pangolins, for example—are arboreal, feeding on the ants and termites that live in the trees.

Perhaps the easiest way to reach the treetops is to fly or glide into them, and the rain forest is filled with mammals that "fly." In South America, in fact, the number of bat species may well exceed that of all other mammal species put together. Many tropical bats are frugivorous (fruit-eating), consuming large quantities of nectar as well as the actual juices and pulp of the fruit. Some of them hollow out places to hang in the clusters of the fruits themselves, while others hang from the tree limbs.

Arboreal rodents do not abound in the South American rain forest, but in Africa flying squirrels are plentiful. And scaly-tailed squirrels have special scales on their tails that help the animals maintain their hold on the branches. In the rain forests of southeast Asia the mammals known as flying lemurs, or colugos, resemble bats in having flight membranes connecting their webbed digits. And in Australia a group of marsupials known as gliding possums are also capable of "flight."

Above: Lush greenery characteristic of the rain forest, Costa Rica. Opposite: Buttressed roots of rain-forest tree, New Guinea

Most of the mammals inhabiting the area between ground and canopy have insectivorous or omnivorous diets. Although foods such as figs and the fruits of the many epiphytes are available on the mid-level plants, it is actually the insectivorous bats that seem most comfortable at this level. Other mid-level feeders include climbers such as the tree shrew, several genera of rodents including tree rats, and in Australia, at least two species of tree kangaroos and a few possums. Several carnivores such as the linsang and various civets also forage over considerable vertical distances in search of food.

Ground feeders in the rain forest fall into one of three categories—the relatively large ungulates; the small to medium-sized rodents, insectivores, and some edentates; and a few terrestrial carnivores. Except for the last group, these mammals, like the canopy ones, are primarily vegetarian. However, the type of vegetation they eat is different, since relatively little fruit from the uppermost layers falls to the forest floor where these mammals can reach it, and that which does decomposes quickly. The larger ground feeders browse on what foliage they can reach, root out underground foodstuffs such as roots and bulbs, and search out numerous fungi. Some of the rodents and shrews climb a bit, and so have access to the food of the lower tree trunks as well as to ground food.

What is particularly interesting about the larger terrestrial mammals is not only that there are very few of them, but also that those species with close relatives in open country are by comparison very small. The situation is perhaps most noticeable in the forest buffalo, the hippopotamus, and the various dwarf antelopes, all of which are diminutive in relation to their savanna cousins. This is no doubt due to several factors, the relative lack of grass in the heavy shade of the rain forest being one of the most important. Another is that a large body size would hinder movement through the forest, particularly if accompanied by the large antlers typical of so many open-country ungulates. The small headgear and generally pointed faces of most rain-forest ungulates are a help in maneuvering through the biome's woody growth.

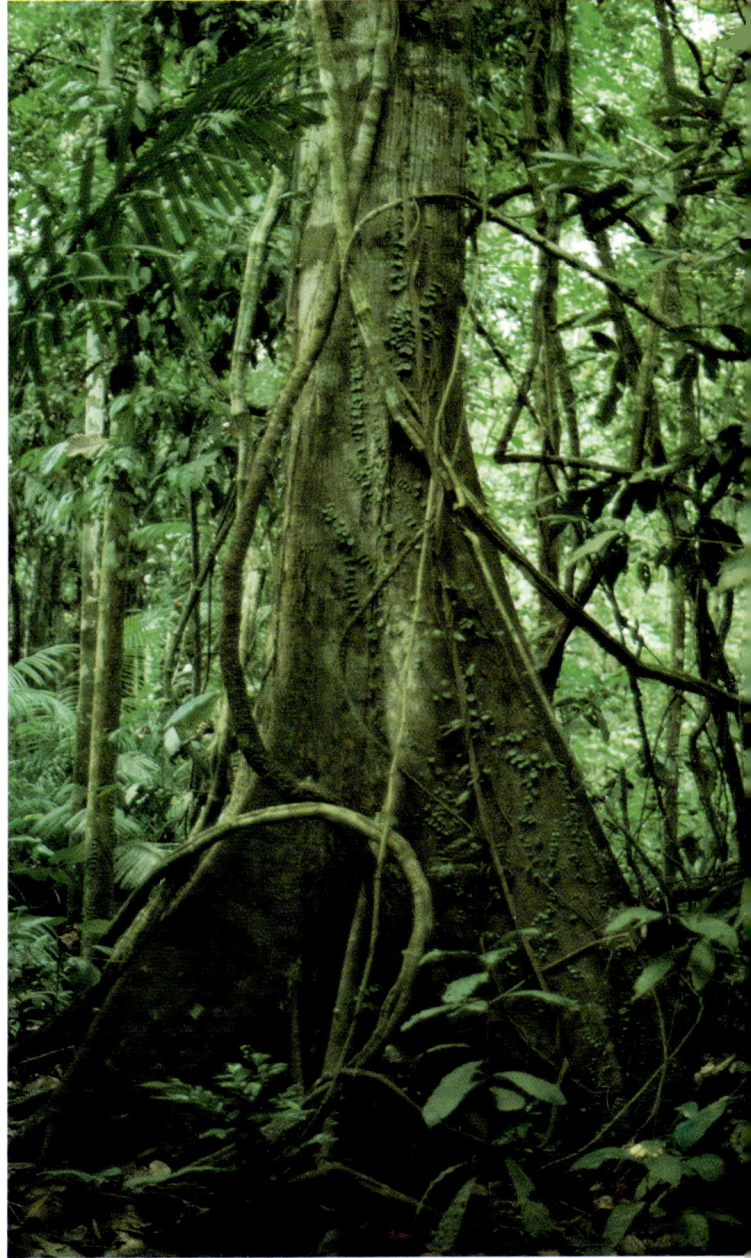

Murine Opossum
Marmosa sp.

Opossums are New World marsupials of the family Didelphidae, not to be confused with the possums of the Australian realm. Most opossums are arboreal, have prehensile tails, and carry their young in pouches. But female murine opossums are pouchless and transport their nine to nineteen poorly developed newborns on exposed nipples, to which the offspring cling tenaciously with their mouths. Specialized nasal passages help the attached newborns to breathe and swallow simultaneously.

These opossums are small, generally weighing less than 3.3 ounces (95 g), but they will not hesitate to attack large grasshoppers, which they kill by biting repeatedly in the head and thorax. They devour all but the very toughest parts of their prey. Various fruits are also included in their diet, especially figs, bananas, and mangoes. Occasionally a murine opossum will turn up in a warehouse somewhere, buried in a bunch of bananas.

Water Opossum, or Yapok
Chironectes minimus

Yapoks are unique in being truly aquatic marsupials. Their webbed feet and thick tail work together to propel their slender, streamlined bodies through freshwater rivers and lakes. Facial bristles help the yapok find its submarine prey of crayfish, shrimp, and other aquatic animals.

The yapok's most distinctive feature is its specialized pouch. Both sexes have a pouch; the male's houses the scrotum, and the female's is used to carry the several young, even in the water. When the female dives, the pouch's rear-facing opening closes by means of strong muscles, forming a watertight compartment that stays closed as long as the mother stays submerged. During this time the young draw their air from the limited supply of oxygen in their closed mini-environment.

These opossums are seen very infrequently, perhaps because they are nocturnal and perhaps also because they spend a great deal of time in their burrows, located along stream banks.

* Haitian Solenodon
Solenodon paradoxus

These New World insectivores are nearing extinction, due to a low rate of reproduction (usually one or two young are born per year), and to heavy predation by mongooses and by domestic dogs and cats that have been introduced into the solenodons' native habitat.

The solenodon is rather amazing in that it is one of the few mammals to possess truly poisonous saliva. After the animal bites its victim, it injects the wound with saliva from an area near the base of the lower incisors.

This mammal uses its long snout as a rooting tool, and the foreclaws are helpful in picking small animals and some vegetable matter from the soil. With some effort the solenodon can sit upright on its haunches by using its nearly naked tail as a prop. It moves along in what can only be described as a waddling gait.

Wrinkle-faced Bat
Centurio senex

Centurio senex is one of the most grotesque and fascinating of all bats. A unique collar of skin covers the bat's face as it roosts, yet apparently permits light to reach the eyes. The collar can be stretched taut near the top of the head. When the bat flies, the collar becomes a large set of wrinkles on the lower part of the face. Males are equipped with additional chin folds, which probably house scent glands. Even this bat's method of feeding is unusual, and not fully understood. *C. senex* has a tiny throat opening, not quite a twentieth of an inch (1.3 mm) in diameter. Somehow it is able to get fruit, its main food, into this opening. It may strain the food through projections located in the area between its gums and its lips.

Wrinkle-faced bats live in trees, roosting under leaves in groups of two or three up to a dozen.

Disk-winged Bat
Thyroptera tricolor

Disk-winged bats get their common name from the adhesive disks attached by short stalks to their wrists and ankles. These disks are bathed in a sticky material, which, along with the bat's strong muscles, can hold the animal upright on a perfectly smooth surface. Each disk alone can support the bat's weight, in fact, and the animal assumes a head-up posture when at rest beside rolled-up leaves or fronds. This position is unusual among chiropters, most of which hang upside-down as they roost.

These bats have an Old World counterpart in the sucker-footed bats of Madagascar, *Myzopoda aurita*. The only bat that inhabits Madagascar exclusively, *Myzopoda* has suckers that are not projected on stalks but are attached directly to the wings. Otherwise their disks are almost identical to those of *Thyroptera*. No one seems to know for sure whether the disks have evolved in these two geographically separated genera as a result of convergent evolution, or whether the two genera are in fact related.

Fisherman Bat, or Mexican Bulldog Bat
Noctilio leporinus

This is one of only three genera of bats known to eat fish. *Noctilio leporinus* most likely detects its prey through echolocation, although it may catch some of its food simply by dragging its feet through the water as it makes its zigzag flights. If echolocation is used, it may serve primarily to pinpoint fish near the water's surface, since most of the bat's ultrasonic frequencies are invalidated by the water. These bats hunt over salt water as well as fresh, and once they locate a fish or crustacean, they lift it out of the water with their large, clawed hind feet. They can swim if they have to, by rowing through the water with their wings. Any prey they do not consume in flight they store in their cheek pouches and bring out later, while they roost.

Fisherman bats are opportunists, and have been known to ally themselves with pelicans, which often panic fish and send them racing off near the water's surface, where they are easily located by the bats. They take some food from the surf, and they eat a certain number of insects as well.

Common Vampire Bat
Desmodus rotundus

Exaggerated legends about vampire bats' blood-sucking have existed for centuries. The real danger posed by this bat's diet and method of feeding, however, is not blood loss to the victim—the amount of blood lost is relatively insignificant—but the readiness with which rabies and other diseases are transmitted through the wounds the bat inflicts. In some countries, in fact, the spread of rabies due to vampire-bat bites is so great that cattle are routinely inoculated against the disease.

One of the reasons these animals are so successful at their particular way of feeding is that they have foot pads that relieve them of having to cling to their victims. And because they can fly, walk, hop, and run over their victims, they often attack, feed, and leave without having been detected. Warm-blooded vertebrates, usually fairly large ones, are their sole victims. The bats choose a hairless spot to puncture with their sharp canines. They make a wound only about a fifth of an inch (4–5 mm) wide, though it oftens seems much larger since anticoagulants in the bats' saliva cause blood to flow freely. Then they lap the blood with their tongues. If the meal is completely satisfying, which it usually is, vampire bats become bloated and do not need to eat again for several days.

Mexican Fruit Bat
Artibeus jamaicensis

The pulp of soft, juicy fruits passes through this bat's digestive system very rapidly, usually in less than half an hour. The indigestible parts of the fruits, such as seeds and cores, are discarded and may accumulate beneath roosting sites, where the seeds sprout and mature. By spreading the seeds in this way, *Artibeus* is instrumental in the distribution of tropical fruit plants.

Artibeus species vary in their manner of homesteading, *A. jamaicensis* being found in tree hollows and caves. At least one *Artibeus* subspecies, *A. cinereus watsoni*, modifies its habitat by cutting across the pleated fan-shaped leaves of various palms. The frond bends at the cut, forming a tentlike shelter for the bats to nestle under. Bats of the genus *Uroderma*, which alter their surroundings in a similar way, are called tent-building bats.

False Vampire Bat
Vampyrum spectrum

The Chiroptera order (bats) contains about 900 species, all but about 130 of which are grouped in the suborder known as the Microchiroptera. This suborder is made up of all bats except members of the family Pteropodidae, which includes all Old World fruit bats. The false vampire bat is among the largest members of this huge suborder, and the biggest of all bats in the New World. It gets its common name from the fact that it was once believed to eat blood, as true vampire bats do. That belief has been disproved, and now the false vampire bat is known to eat rodents, birds, and even small bats—a truly carnivorous diet, rare among the chiropters.

False vampire bats belong to the Phyllostomatidae family, the American leaf-nosed bats. The shieldlike flap of skin that appears to be stuck onto the end of this bat's nose is thought by some scientists to influence the scope of the beam of sound that the animals emit during echolocation.

* Golden Lion Marmoset
Leontideus rosalia

Marmosets and tamarins together constitute the family of New World primates known as the Callithricidae. They are important to an understanding of evolution because they are so much more primitive than the other New World monkeys, which are often lumped together in the family Cebidae. For example, marmosets and tamarins have clawlike digits instead of nails, a situation that makes gripping very difficult. Because of this, they tend to climb vertically, much like a squirrel, by digging the claws into tree bark. Tamarins and marmosets typically have multiple births, also a feature of more primitive species (although the meticulous caring for the infants that the fathers exhibit is not). And the brains of these species are relatively undeveloped.

The pygmy marmoset, *Cebuella pygmaea*, is the tiniest of all living monkeys. It feeds in a somewhat unusual way. Using its specially adapted lower incisors, it drills shallow holes in tree trunks. Then it removes and eats the sap, which is probably the most important element of its diet. It supplements the sap with berries and insects.

It is more than unfortunate that the members of this rather unusual primate family are quickly being wiped out. Only several hundred golden lion marmosets remain in the Brazilian rain forest, primarily because the cultivation of land there has destroyed much of their habitat. The cotton-top tamarin, a rather fragile-looking species, has recently joined the golden lion marmoset on the endangered list.

▲ *Golden lion marmoset* ▼ *Spider monkey*

Spider Monkey
Ateles sp.

Spider monkeys could not be more appropriately named. Their tails are so prehensile that they actually function as a ''bonus'' appendage, creating the impression that the animal has more than four limbs. Furthermore, when spider monkeys run along a branch on all fours, the joints in their very long arms and their legs angle up in a spidery fashion. The entire locomotion process is an exercise in the superb coordination of many body parts. In addition to being able to run with ease, these mammals can stand and walk upright, can clear many feet in a single leap, and can brachiate on their thumbless hands.

Spider monkeys are perhaps the most adaptable of primates, and are even capable of varying the size of their social units in response to food availability and other external conditions. Unlike the squirrel monkeys, spider monkeys are found alone from time to time, particularly the males, which may dissociate themselves periodically from the bands of females and young.

Capuchin
Cebus capucinus

Capuchins are not only the most common of all New World primates, they are also among the most adept at using their hands, which have opposable thumbs. Because of this adaptation, they are able to enjoy a varied diet. These monkeys seem to have a compulsion for taking things apart, not only nuts and oysters (which they open by pounding with a rock or against a hard surface), but also rolled-up leaves containing larvae, bark concealing insects, and nearly anything else they can poke at. But their attention span is short, and if they fail to open an object quickly, they move on to something else. Their habit of biting into fruits and then dropping them uneaten may speed ripening, and these monkeys are known to return later on and eat the fruits they had dropped earlier.

These primates' need to manipulate seems to extend even to their use of their tails. They often use their tails instead of their hands—to carry food, for example. And some captive individuals have been observed catching objects thrown to them, using the tip of the tail as frequently as the hands. For just such reasons, these spirited monkeys make fascinating pets. They are the traditional organ-grinders' partners.

* Red Uakari
Cacajao rubicundus

The uakari is the only New World monkey without a long tail. Normally the lack of a long tail would indicate a ground-dwelling life style, but this is far from being the case with the uakari. Uakaris rarely touch the ground, but leap about in the top levels of the rain forest.

A rather remarkable combination of features gives the red-faced species its bizarre appearance: rosy skin that takes on a deep blush when the animal is excited, a bald skull-like head conformation (due to the absence of fat), and uneven streamers of red hair. When the animals run, their light-colored flesh is highlighted by the billowing hair.

Uakaris and their close relatives the sakis illustrate why the New World monkeys are known as Platyrrhini, which means "broad-nosed." Their nostrils are rounded and set far apart, in contrast to the Old World monkeys' downward-pointing ones. Both uakaris and sakis have protruding lower incisors, which are thought to be used for impaling fruits. (These primates also eat foliage, buds, and seeds.) Remarkably little is known of the uakari's breeding habits, since the species has never been bred in captivity.

Howler Monkey
Alouatta sp.

At 15 to 20 pounds (7–9 kg), the howler monkeys are the heaviest of all the nonhuman New World primates. Their rather simple digestive systems can process young, tender leaves as well as fruits, and howlers forage along well-established "highways" of branches stable enough to support their weight. They are not brachiators, but usually move on all fours on top of the limbs. The first and second digits of their hands are opposable to the other three, a feature that facilitates grasping and climbing. They use their prehensile tails as extra support during frolics or periodic upside-down feedings.

Howlers produce what may be the most distinctive sound in the mammal world, a roar so loud that it carries for several miles. The roar is usually initiated by the old males that lead the four- to forty-member troupes, then is sometimes picked up by the other troupe members. The old males not only have enlarged resonating chambers, they also have long "beards," the loudest species having the longest beards of all. The purpose of the vocalization seems to be to alert other troupes of their whereabouts and thus avoid conflicts over feeding territories. Roaring is more frequent in areas with high howler populations than it is in less crowded areas of the rain forest.

Night Monkey, or Douroucouli
Aotus trivirgatus

The night monkey, or douroucouli, is the only nocturnal monkey. In fact, it is the only nocturnal primate of any sort found in the New World. Its large eyes are indicative of its nighttime activities. The eyes have a transparent cornea that aids in the collection of light just before dawn and after sunset, when douroucoulis are most active. Apparently the cornea allows the animal to see well enough to capture flying insects and even to bound through branches. The douroucouli's long tail helps it move efficiently, acting both as a balancer and as a brake.

During the day, the douroucouli sleeps, along with its mate, in a hole in a tree. It never wanders far, and returns to the same nest each morning. This habit, which limits the feeding territory to a relatively small area around the permanent home, is very different from the nomadic life style of most higher primates.

Douroucoulis' fur is short and thick; its texture may serve some adaptive purpose, such as muffling body-movement sounds. Many other nocturnal arboreal mammals of the New World have a similar coat. Douroucoulis produce a wide variety of noises, from a loud boom to a delicate twitter.

Squirrel Monkey
Saimiri sciureus

Squirrel monkeys, perhaps the most social of all New World primates, are rarely found without companions. Although they exhibit some irritation at being crowded, they will tolerate poor habitat and other inconveniences just to avoid isolation. Under most circumstances individuals are in sight of at least one or two other squirrel monkeys, and will join readily with certain types of capuchin monkeys if left behind on their own.

There is a certain amount of fluidity in total group size among squirrel monkeys. Up to five or six hundred squirrel monkeys have been reported to move together through the trees, and the species is known to gather in considerable numbers for sleeping. It may be that when quantities of food are available in a given area, the larger troupes forage together, then break into smaller feeding units at other times.

Ecologically these monkeys are described as diurnal arboreal omnivores, although they actually range from treetop to ground in their search for food, scrambling and leaping through the forest with great agility. They seem to prefer areas near riverbanks. Their tails, though not prehensile, are an effective balancing aid, and their hands adhere to nearly all types of surfaces.

Giant Armadillo
Priodontes giganteus

South America's giant armadillos and Africa's terrestrial pangolins have amazingly similar bodies. Both are covered with protective scales, the pangolins' being made of compressed hair, the armadillos' of skin-covered bone; both have sharp, somewhat backward-pointing claws; and both have long extensible tongues well suited for capturing insects (though armadillos are not as strictly limited to that diet as pangolins are). It is no surprise, therefore, that taxonomists originally classified the two animals in the same order, Edentata. More recently, they have concluded that the similarities are the result of convergent evolution, not of real kinship, and the pangolins have been assigned their own order.

Giant armadillos can weigh as much as 130 pounds (59 kg). Although Edentata means ''without teeth, ' it is a misleading name, for this species has as many as a hundred teeth when young, then sheds them as it ages. These animals reportedly supplement their rather generalized diet of insects, snakes, larvae, and worms by feeding on newly interred human corpses.

Giant Anteater
Myrmecophaga tridactyla

Anteaters, the only living toothless mammals, are equipped for an insectivorous niche by their rather miraculous tongues. After their strong claws have broken open an ant or termite nest, the anteaters extend their elongated snouts deep into the nest. Their cylindrical tongue, coated with sticky saliva, is capable of trapping up to thirty thousand insects in a single day. An extraordinary amount of saliva is necessary to accomplish this feat, and the giant anteater's salivary glands extend all the way from the sternum to the tongue.

The claws and feet of the three anteater species provide clues to their respective habitats. Giant anteaters' foreclaws turn under, and the animals walk and even gallop on their knuckles. These anteaters do not climb trees. The tamandua and the pygmy silky anteater are both arboreal, and they walk on the sides of their forefeet with their foreclaws turned sideways. Both arboreal species have a prehensile tail, which they wrap around tree limbs and with which they brace themselves when they assume a forward-facing defense posture. When the animals are in this position, the tail and hind feet form a tripod on the branch, leaving the forearms free to strike out at enemies. Were it not for their strong forearms, these slow, toothless anteaters would be defenseless.

▲ Giant anteater

▲ Two-toed sloth

Two-toed Sloth
Choloepus sp.

Its upside-down position is the sloth's most striking feature. Sleeping, eating, mating, even giving birth are all carried out from this position. The sloth's long appendages are very strong, and its claws can support the animal's entire body weight when they are hooked over the limb of a tree. (Sloths of the genus *Bradypus* have three foreclaws, or toes, and nine neck vertebrae; sloths of the genus *Choloepus* have only two foreclaws and six neck vertebrae.) When the animal is upside-down its fur hangs down, thereby shedding rain water. During the heaviest rains, the fur hosts an algal growth that gives the animal a greenish tinge. This coloration provides effective camouflage, for it makes the sloth look like a bunch of leaves.

Sloths browse somewhat lethargically for tree fruits and leaves, moving along hand-over-hand between feeding areas. (They are practically helpless on the ground.) Their slowness is due not to laziness, as the common name implies, but to a very low metabolic rate and a small heart. Even their sluggishness has survival value, though, for it makes the animals inconspicuous in the canopy.

Since the warm-bloodedness of these mammals is imperfectly regulated—their body temperature may fluctuate as much as 15 or 20° F (10–11° C)—sloths would find it difficult to survive in any climate but a tropical one.

Capybara
Hydrochoerus hydrochaeris

At 110 pounds (50 kg) or more, the capybaras are the world's largest rodents. They are so large, in fact, that they might be mistaken for pygmy hippos if both animals lived on the same continent. The similarities between these two totally unrelated species are due to convergent evolution, and in this case go beyond physical appearance to include certain other adaptations such as the alignment of eyes, nose, and ears above the surface of the water.

As evidenced by its partly webbed feet (as well as by its Latin name, which means "water hog"), the capybara spends much of its time in the water. It eats a wide assortment of aquatic plants while standing partly submerged, and it can swim underwater long enough to escape from most enemies, which include jaguars and caymans.

This species has a calm temperament, which is reflected in its social life—there is little infighting among the three to thirty herd members. Capybaras seem to have a rather intricate system of communicating with one another, using a large repertory of sounds ranging from grunts and barks to clicks and whistles. The herds do not make dens, but rest in shallow depressions in the ground.

Prehensile-tailed Porcupine
Coendou prehensilis

Prehensile-tailed mammals are fairly common in the New World. But the prehensile-tailed porcupine is unique among porcupines in being able to grip with the upper side of its tail rather than the under side. This means that it wraps its tail around branches in what can only be described as an "underhand" position, getting its real grasping power from the naked area near the tail's tip. This porcupine climbs very deliberately and sure-footedly from limb to limb. In fact, its slowness, like the sloth's, may provide camouflage, since the porcupine is easily mistaken for a bunch of tangled vegetation. Usually it reaches for the branch it is moving toward with its well-clawed forefeet, then releases its grip on the original branch and swings its weight under it. With its tail and hind legs, it grabs the branch it is going to, then climbs aboard.

The quills of these (and other) porcupines are armed with barbed hooks that make them deadly to enemies. They cover most of the animal's body, except for the tail, feet, and nose.

Prehensile-tailed porcupines are nocturnal and, as one might guess, arboreal. They spend their days asleep in the tangle of greenery near the tops of trees, or sometimes enter a hollow trunk for shelter.

Paca
Agouti paca

The paca is unique in the mammal world because of its particular sound-producing abilities. A rumbling noise is produced as the paca blows air from one set of cheek pouches into another, while a specialized chamber provides resonance.

The paca is by no means an aquatic species, yet it often escapes its enemies through water and lives in burrows along riverbanks, as well as in tree roots and under stones. It spends the day "indoors" and emerges in the evening to feed on mangoes, avocados, and other vegetable delicacies. Its flesh is considered such a culinary treat by many native people that specially trained dogs are often used to search the rodents out.

Several New World rodents of the cavy, or guinea pig, type bear striking similarities to Old World ungulates. The agouti, for example, shows a marked resemblance to the royal antelope, and the paca to the African chevrotain. These resemblances are the result of convergent evolution.

Jaguar
Panthera onca

Jaguars, jaguarondis, margays, tiger cats, and ocelots are all members of the cat family that are found in the South American tropics. Of these the jaguar is by far the largest and most formidable, weighing well over 250 pounds (93 kg) when full-grown. This is larger than both leopards and mountain lions, and is nearer the size of a lion or tiger.

Although jaguars will take an occasional sloth in the trees, their size limits climbing ability, and they spend most of their lives on the ground. They hunt along streams in the rain forest, where they fish expertly with their powerful claws and wait for animals that come to drink. They prey heavily on capybaras and peccaries, and occasionally hunt large domestic species, including cows. They may also try to attack giant anteaters, but are often unsuccessful—jaguars have been known to be killed by the anteaters' powerful claws.

Once it has brought down a small victim, the jaguar eats it quickly in its entirety. Large prey is chewed on for a while where it is brought down, then is dragged to a secluded spot where it will be safe from vultures.

Tayra
Eira barbara

Tayras, members of the mustelid family, are the South American equivalents of the northern hemisphere's martens. Both terrestrial and arboreal, they are extremely fierce, and will attack birds, a variety of small rodents, and even brocket deer, though their diet also includes honey and various fruits.

One of the tayras' most effective escape tactics, often used to outwit hunting dogs, is to speed through the trees and run along the ground alternately. Tayras are agile even in the water, and may pursue their prey across a pond. They hunt in groups as well as singly, and have been known to do serious damage in banana plantations, where they seem to relish the ripened fruit.

Little is known of the breeding habits of this carnivore, but the male's baculum—the bone that reinforces the penis in many mammals—supposedly has aphrodisiac qualities. In some parts of Mexico the animal is given the common name *cabeza de viejo*, which means "head of an old man." No doubt this refers to the tayra's gray throat and head area, which are in dramatic contrast to its darker body.

Kinkajou
Potos flavus

These are the only members of the raccoon family with a prehensile tail. They use the tail to "buy time" between footholds in the branches—that is, they will not release their tailhold on a branch until their feet are firmly planted in a new spot. They do not grasp food with the tail, as monkeys do, but they are very monkeylike in other respects and in fact fill an arboreal niche similar to that of the New World primates. They behave particularly like night monkeys (douroucoulis) in that both animals are thought to return each morning to their permanent homes (usually tree hollows). They sleep throughout most of the day, coiled up with their front feet over their eyes. Occasionally kinkajous will emerge before nightfall and drape themselves over a mass of vines.

Olingos, their close cousins, often forage in parties with kinkajous. Fruits and insects form the bulk of both animals' diets, although the kinkajou has a long tongue well adapted for extracting honey from a hive. In spite of their mixed feeding habits, both animals are classified as carnivores.

Coatimundi
Nasua nasua

Coatimundis, or coatis, look like no other animal on earth, but appear to be a composite of several. In their masked faces and bushy ringed tails one can read their family tree —the raccoons and their relatives. Yet ring-tailed lemurs have similar tails, and even hold them up like plumes just as coatis do when they feed. The coati's tapered and pointed snout looks very doglike and is an effective probe for poking about in holes. Coatis are known to roll their prey under their feet, which rids it of its spines or (in the case of biting or stinging species) kills it.

One fascinating aspect of the coati's life style is that it is a truly communal animal. Coatis have evolved elaborate vocalizations and social gestures, including mutual grooming, nose rubbing, and nuzzling, and they may be as sophisticated in their use of these signals as are chimpanzees and other primates. Such communication is rare for this type of carnivore. Older males are sometimes found in separate parties from the females and young.

Interestingly, coatis are extending their range northward, and are well adapted to mixed habitat with some grassland as well as to the rain forest.

* Central American Tapir
Tapirus bairdi

This relative of the horse and the rhinoceros is among the more primitive of larger mammals, as evidenced by its low, plump shape, its unspecialized teeth, and other anatomical features. Its trunk serves as a sensitive olfactory organ, and the bristles at its tip are tactile organs with which the tapir explores its surroundings. It also uses the trunk to pluck fruits, grasses, leaves, and even aquatic plants.

Water is an important part of this shy mammal's world, and wallowing in mud is a major means of controlling insect pests. When not disturbed, the tapir stays on well-defined pathways it has created itself, most of which lead to a river or stream. If it is threatened by a jaguar or some other enemy, however, the tapir leaves its pathways and tears through the brush.

One of the four species of tapirs is found in Malaya; the other three inhabit Central and South America. All but one of the four are on the endangered list.

▲ *Coatimundi* ▼ *Brocket deer*

Brocket Deer
Mazama sp.

The brocket deer of the New World and the duikers of the Old World occupy the same niche, that of terrestrial browsers and grazers, and both types of ungulates are well adapted for it. Their rather small body size allows them to move easily through thickets and tangles of greenery, and their pointed faces and short, simple antlers, just a few inches long, do not impede movement. (There seems to be no fixed season for shedding the antlers.) It has been suggested that the brocket deer's ancestors were true savanna species with elaborate antlers of up to eight points, and that the size of the antlers became reduced as the deer gradually moved into the forest.

Brockets are shy mammals that spend much of their day secluded under fallen tree trunks or between large roots. They come to the forest edge at dusk and often disturb domestic crops in their search for food. Because they freeze when they sense impending danger, they are often able to escape detection and attack by people, dogs, and other enemies such as jaguars.

Giant African Water Shrew
Potamogale velox

Giant African water shrews are large insectivores that have become adapted to an aquatic life style. Also called otter shrews, they resemble otters in their long, lithe bodies and thick tails, and like many otter species they hunt fish and other aquatic animals. Crabs, a favorite food, are brought ashore and ripped open with a quick bite or two.

Despite their aquatic life style, these shrews have no semblance of webbed feet; they seem to propel themselves through the water solely through movement of their laterally flattened tail. When they swim, otter shrews hold their hind feet very close to the body, thereby increasing streamlining. A special flap prevents water from flowing into the nostrils when the animal submerges. Otter shrews make their homes in holes in riverbanks, with the entrances well below the water level.

The giant African water shrew is an expert climber and can twist itself around well enough to bite an enemy that is holding it by the tail. Its bite is powerful, and only with great effort can the water shrew be forced to release its hold.

Hero Shrew
Scutisorex congicus

Hero shrews have a unique vertebral column. The individual vertebrae have interlocking spines as well as dorsal and ventral spines. Together the structures create an exceptionally sturdy backbone. Adult men reportedly have stood on the backs of these shrews for several minutes at a time, with the shrews suffering no ill effects—when the men stepped down, the shrews merely shook themselves and scurried away. One would think that a spinal column strong enough to protect the shrew's internal organs under such weight would be stiff and restrictive of movement, but the hero shrew does not appear to have any difficulty flexing its back. Many natives of the African Congo believe that, because of its extraordinary strength, the hero shrew possesses magical powers, and they believe that they can acquire these powers themselves if they eat the meat of the animal or wear parts of its body. No one really understands the adaptive value of such strength in the shrew, however.

S. congicus eats small animals (and possibly some vegetable food), for which it searches along the ground. Unlike many other shrew species, it does not hunt in an irregular, stop-and-start fashion, but conducts its search in smooth, flowing motions.

Old World Leaf-nosed Bat
Hipposideros caffer

Old World bats of the Hipposideridae family and those of the Rhinolophidae family are closely related, and the females of both families are unusual in having fake nipples. Most bats are born with well-developed teeth, with which they cling directly to the fur of the mother when she is in flight. The young of these two families, however, cling to these "nipples," which are actually just protrusions on each side of the mother's abdomen. They are not connected to mammary glands, and so produce no milk.

Members of these two families also share another characteristic—they wiggle their ears in a predictable pattern during echolocation. The flap of the ear, or ear pinna, moves forward as one high-frequency pulse is emitted, then back as the next one is emitted. The other pinna moves in opposite time. This alternation may help the bat to estimate distances or determine directions.

H. caffer is the only bat that catches insects in the undergrowth of northeastern Gabon's mature forests. Perhaps this unique feeding niche accounts for the dense populations of this bat in that area.

Galago, or Bush Baby
Galago senegalensis

These prosimians were among the earlier primates to evolve. They are kangaroolike leapers with hands and feet adapted for landing and maneuvering on nearly any type of surface. Their extremities have five digits equipped with padding, and their tarsal bones (bones in the heel and ankle region) are extremely long, an adaptation that provides the leverage that enables galagos to make 30-foot (9-m) leaps. Galagos catch flies on the wing by clapping their hands together on the insects and then simply folding their fingers around their prey. They are known to moisten their extremities with urine, probably to give them extra gripping power and to mark territories.

When alarmed or hurt, this animal cries loudly, much as a human infant does. It is not as advanced socially as the primates that groom each other in a ritualized fashion. But it is equipped with a special grooming protuberance under its tongue, which removes foreign matter from its own teeth after the fur has been passed through them.

* Indri
Indri indri

Prosimians are the primitive forerunners of the true monkeys and apes. They are usually classified in one of four categories—tree shrews, lemurs, lorises, and tarsiers. They typically have a keen sense of smell and hairs on their pointed muzzles that act as sensitive tactile organs. Most are nocturnal, have multiple births, mature quickly, and have simple social lives. The lemurs are probably the best-known prosimians, since they evolved on Madagascar in isolation and so radiated out to fill niches occupied by other types of mammals elsewhere.

Lemurs of the family Indridae occupy a niche similar to that of the leaf-eating monkeys of the Asian and African tropics and the howler monkeys of the New World tropics. They are the largest of all prosimians (the indri is 4 feet [1.2 m] long). Their facial conformations, particularly their flattened muzzles, give them a more monkey-like appearance than the other lemurs. Indris have short tails, but their long opposable toe helps them to grasp branches and move through the trees easily. Indris rarely come to the ground. When they do, they walk upright.

Indris have traditionally been unmolested, but are endangered today because of severe habitat destruction.

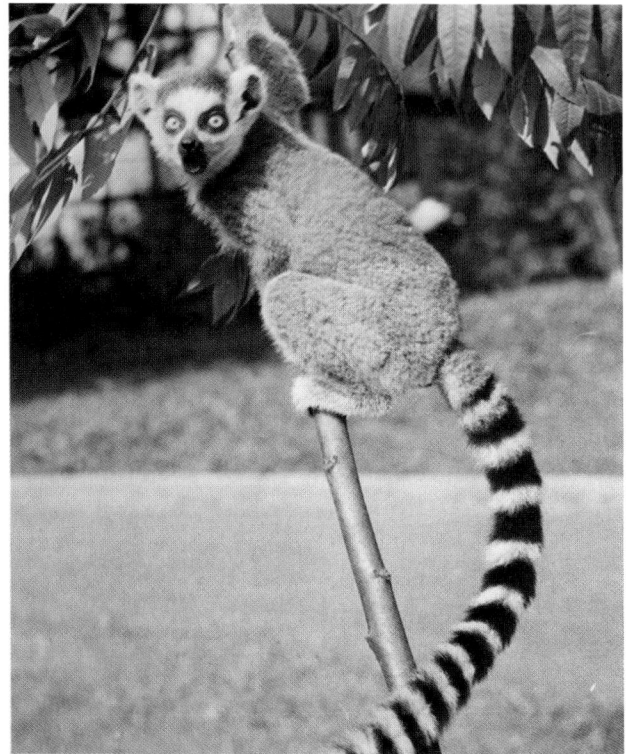

Fat-tailed Dwarf Lemur
Cheirogaleus medius

Fat-tailed dwarf lemurs are so named because they store fat at the base of their tails. When food is scarce they live off this stored fat; they also estivate and thus delay having to satisfy their need for fresh foods.

These tiny animals are entirely nocturnal. They spend the day in treetop nests or in holes in tree trunks. Studies indicate that they may even use birds' nests as their homes. Dwarf lemurs and mouse lemurs resemble the lorises and bush babies much more than they do the large monkeylike indris and lemurs. Dwarf lemurs, for example, have padding on the ends of their digits, as certain bush babies do. There is even a grooming claw on the second toe very similar to that of the slow loris.

The aye-aye, the indri, the fat-tailed dwarf lemur, and the ring-tailed lemur are all lemuroids, a group of prosimians that forms the core of Madagascar's unique fauna. Scientists think that their forerunners reached Madagascar and neighboring islands from Africa, and were able to thrive there in the absence of competition from monkeys. Though not all lemuroids are true rain-forest species, the four treated here have been grouped together as such for purposes of cohesion.

Ring-tailed Lemur
Lemur catta

Whether it is dangling from a branch or held upright in the shape of a question mark, this lemur's tail is by far its most noticeable feature. The tail rings, visible even in dim light, are a clearly identifying mark, and the tail itself is used as a balancing aid as the lemur leaps among the branches. One type of scent marking is accomplished when the lemur raises its tail and defecates or urinates. This action, in addition to being a means of elimination, is used as a warning to intruders.

Ring-tailed lemurs follow a definite daily schedule. They rise early and lick the morning dew from leaves and from their fingers, which they wet by dipping them into tree hollows. After a period of sunning, they move across parts of their rather large territories in one or two waves, feeding on plant foods as they go. (Ring-tailed lemurs differ from most other lemur species in that they travel primarily on the ground instead of in the trees.) The lemurs nap until their next feeding session, in the late afternoon, and sleep in the trees at night.

Troupes, composed of between five and twenty-two members, are dominated by females, though either sex may take the lead as they wander. Females and young always have the first choice of foods, an adaptation that may ensure their survival during times of drought.

Aye-aye
Daubentonia madagascariensis

The aye-aye puzzled taxonomists for years. Now it is considered a prosimian and therefore a primate, but when the animal was discovered in 1780 scientists considered it a type of squirrel because of the structure of its constantly growing incisors and its other teeth. The aye-aye's appendages are also rather rodentlike, the limbs bearing claws instead of nails on all but the big toe. The big toe, however, bears a flat nail; and because the toe is opposable, the feet can grasp branches easily. These two characteristics, plus certain skull features, indicate that the aye-aye is a primate.

With the needlelike digit on the forelimb, the aye-aye taps for larvae in wood, then extricates them by hooking or impaling them. (These claws are also used in grooming the fur and teeth.) Aye-ayes have an acute sense of hearing, and can locate larvae tunnels under bark simply by listening. And their eyes are well adapted for night vision. These adaptations are invaluable to the aye-aye as it moves about in the forests of Madagascar.

Many superstitions surround the aye-aye, and it has been unmolested in most places. However, it is highly endangered today, due to destruction of its forest habitat.

Colobus Monkey
Colobus sp.

Like langurs, colobus monkeys are arboreal leaf-eaters. They must consume large amounts of leaves to get sufficient nourishment, and they have very large stomachs and specialized digestive systems that permit them to do so. The advantage of this rather unusual diet for a monkey, of course, is that it eliminates competition with the fruit-eating primates. Colobus monkeys have only stumps for thumbs and grasp leaves between the fingers and the palm.

Colobus monkeys use their long limbs as they crash and leap from branch to branch. The black-and-white guereza, no doubt the most dramatically marked of colobus monkeys, has a fringe of long white body hair that often helps the animal control its speed, since it catches the wind much as a parachute does. Although a quick escape is their more usual method of defense, guerezas will sometimes remain quite still in the branches. Then the black coloration of these monkeys blends with the foliage, and their long white hair resembles tropical mosses and lichens draped through the branches.

Collared Mangabey
Cercocebus torquatus

These long-tailed monkeys have pronounced white eye-lids and light-toned areas above the eyes. Since collared mangabeys do not have well-developed vocal organs, these markings serve a definite communicative purpose. They stand out even in the darkest part of the forest. Once one mangabey catches another's attention with its eye markings, it communicates by blinking its eyelids very fast in an obviously well-understood set of signals.

These mangabeys occupy the lower strata of forest trees. They often move about on the ground as well, in search of the various fruits and seeds that compose their diet. In this way they avoid competition with those species restricted to the canopy.

Mangabeys have definite rump pads, called ischial callosities, that are typical of Old World primates. These are actually built-in seat cushions that help the mangabeys rest and sleep while sitting up in a tree or on the ground. Attached directly to bones, they contain no blood vessels or nerves and therefore can be subjected to several hours of pressure without causing the animal discomfort. Ischial callosities are absent in New World primates, which are more arboreal.

* Gorilla
Gorilla gorilla

The gorilla, a vegetarian and the largest living primate, lives in well-organized troupes, each led by a silver-backed male. The entire sixteen- to thirty-member troupe follows this male's lead in all decisions, from when to get up in the morning to where to build their nests at night. By facing a particular direction within the animals' 15-square-mile (39-sq-km) range, the dominant male indicates where the troupe will begin to forage. And he defends his troupe with an impressive display of chest-beating, hooting, and tearing at vegetation. Apart from these displays, the troupes are silent—since they are almost constantly together, they have no need for elaborate vocalizations to locate one another.

Because of their size, up to 450 pounds (204 kg), the largest adult gorillas are not adapted to climb trees, and usually nest on or near the ground. The smaller females and juveniles build their nests in the treetops. The gorilla's extremely long arms indicate, however, that at one time these primates, or at least their ancestors, were arboreal. Today the gorilla walks on all fours, although it can stand and walk upright for short distances.

Lowland gorillas are found in western and central Africa, mountain gorillas in east-central Africa only. The latter group is highly endangered.

Mandrill
Papio sphinx

Gaudy coloration marks the male mandrill fore and aft. His light blue and deep red facial markings are only slightly more dramatic than the pinkish to lavender coloring on his buttocks and genitals. Females, on the other hand, are relatively drab. This extreme difference in appearance between sexes of the same species (known as sexual dimorphism) is particularly noticeable in Old World primates, as are marked differences in size and dentition. There are many exceptions to it, but the theory that best explains such differences is based on the extent to which a species is specialized for terrestrial living—the more time spent on the ground, the greater the need for male dominance and protection of the females and young, and therefore the greater the physical differences between the sexes. What is interesting in the case of the mandrills (which do feed on the ground and live in male-dominated troupes) is that their close cousins the drills are comparatively bland in appearance; yet the two have similar life styles, and the drill seems at no disadvantage.

Mandrills and drills are omnivorous, eating fruits, roots, snails, insects, and even some amphibians and reptiles. Occasionally they eat small rodents as well.

Chimpanzee
Chimpansee troglodytes

Chimpanzees have been the subject of so much research that by now they are well known for their superior intelligence. What may not be so well known is their capacity for making and using tools in the wild. Chimps will thrust a stick into a termite mound, for the insects to cling to, then bring it out and eat the termites off. If a stick is too large, the chimp will strip its leaves off before inserting it. Such behavior seems to place chimpanzees, of all primates, the closest to human beings.

Advanced social and emotional responses also characterize the anthropoids, the group of apes that includes chimpanzees, orang-utans, gorillas, gibbons, and siamangs. Chimps live in groups of from twenty-five to eighty members, and the range of communicative signals they use with one another is extremely complex. Social stratification, however, is not as rigid as it is in baboon and gorilla troupes—individual males gain and relinquish dominance rapidly. Females generally mate with several males, and bonds between mothers and young are visibly strong.

Chimpanzees typically eat fruit, but occasionally a group of them will kill an animal in order to obtain meat. They are both arboreal and terrestrial and, like gorillas, construct new sleeping nests each night.

Pangolin
Manis sp.

The tongue, stomach, tail, legs, and body covering of the pangolin are all highly specialized. The tongue extends all the way from the pelvic region, where it is anchored, to the mouth. Covered with sticky saliva, it shoots out of the pangolin's mouth to capture ants and termites; it laps up water with a repeated in-and-out motion. The pangolin's stomach is equipped with horny protuberances that grind the food.

Some species are arboreal and some terrestrial, but nearly all pangolins must use their tail as a brace in order to walk on their hind feet. The tail of the arboreal species is a wonderful multifunctional aid—it can support the entire weight of the animal as it hangs from a limb, and it can also be twitched to scare off predators. One of this mammal's best defenses is to curl itself into an impenetrable ball of scales; another is to eject a foul-smelling liquid from its anal region.

The shape of the scales varies from species to species, and sometimes even within a species. The tails of the five arboreal species have a naked area on their underside, while those of.the two terrestrial species do not. This scaleless area, however, does not necessarily function as a tactile organ, as one might expect it to.

West African Scaly-tailed Squirrel
Anomalurops beecrofti

There are few arboreal rodents and no flying squirrels at all in the New World tropics. The Asian tropics are replete with flying squirrels. And Africa has a unique family of arboreal rodents known as the scaly-tailed squirrels, which are not true flying squirrels but have evolved similar adaptations as a response to nearly identical environmental pressures. The members of three of the four genera are equipped with a gliding membrane that is supported by a length of cartilage extending from the elbow joint. (In true flying squirrels this membrane is attached at or near the wrist.) In all scaly-tailed squirrels, the underside of the tail has scales near its base. These serve to prevent skids when the animal lands on a tree trunk, and they are also an aid to climbing.

Scaly-tailed squirrels feed on bark, on some insects, and on nearly all the soft parts of trees, such as the fruits. *Anomalurops* even eats the palatable layer under the bark.

▲ *Chimpanzee* ▼ *African dormouse*

African Dormouse
Graphiurus murinus

Dormice look somewhat like miniature chipmunks, and all but one of the seven genera inhabit the Palearctic realm. The African dormouse is classified in that one genus. Oddly enough, at least some of its twenty or so species become dormant from autumn until spring, just as their temperate-zone relatives do. In preparation for their dormancy they put on considerable body fat, then curl themselves into a circle to sleep. In some parts of the African dormouse's range, the young are born during the period of dormancy.

Graphiurus is often quite a pest, since it raids domestic poultry yards and may make its home in upholstered furniture. In the darkest part of the rain forest African dormice are sometimes diurnal, though they are usually active at night. The meat they occasionally eat is balanced with foods more typically eaten by rodents—fruits, seeds, and nuts. Some of this food is stored for the eating "intermissions" the dormice take during their dormant period.

Oil-palm Squirrel
Protoxerus stangeri

The oil-palm trees that grow over much of tropical central Africa bear orange-colored nuts that form part of this tree squirrel's diet. It is from this tree that the rodent gets its common name, and from the nuts that its pelage often becomes stained. Because oil-palm squirrels also gnaw on bones and tusks, they are sometimes called ivory eaters.

This is one of the largest of all tree squirrels, measuring some 20 inches (50 cm) from its head to the tip of its tail. Little is known of its habits except that it makes its home in tree hollows and feeds both in the branches and on the ground. *P. stangeri*, one of two *Protoxerus* species, produces a booming call when disturbed, and is consequently called a boomer. At other times it may emit twitters reminiscent of bird language.

African Palm Civet
Nandinia binotata

The Viverridae family includes the civets, the linsangs, the genets, the mongooses, and the fossas. These are the typical small- to medium-sized carnivores of the Old World. Some of them are remarkably catlike in the way they lift their paws, sit hunched over their forefeet, and move with exceptional grace. In fact, some taxonomists classify the fossas with the cats.

The civets bear the name of the strong-smelling anal secretion that they and most other members of the family discharge when threatened or when marking territories. This fluid has been used for many years in the manufacture of perfumes, and it also has some medicinal value.

African palm civets are nocturnal and are active both in the treetops and on the ground. In addition to eating large amounts of fruits, they prey on pottos and rodents. They are easily tamed and have been used to keep rats, mice, and cockroaches from overrunning private homes.

Golden Cat
Profelis aurata

Golden cats are among the least known of all felines. One species inhabits the African rain forest, a second inhabits Asia, and a third, known as the bay cat, inhabits Borneo. That three very similar forms should have such scattered ranges suggests that the golden cat originated in one area, then migrated to the others.

About twice the size of domestic cats, African golden cats vary widely in coloring, from a deep chestnut to a slate gray. Some individuals are heavily spotted; others have spots only on their abdomen and limbs. Many African tribespeople believe that golden cats' skins have mystical powers—some say they assure good elephant hunting, and others have long used them in their ceremonial dress. Killing golden cats for tribal purposes is often a matter of strict secrecy.

Profelis aurata is a nocturnal hunter capable of bringing down a wide variety of prey, from domestic poultry to forest ungulates. It usually spends the daytime hours in its den, which may be in a tree or in dense undergrowth.

Tree Hyrax
Dendrohyrax sp.

Tree hyraxes are browsers of the canopy. Their very specialized feet help them cling to trunks and branches. The soles of the feet form a partial vacuum each time the animal takes a step, thereby "gluing" the hyrax to smooth tree bark.

These animals are very vocal, and in fact are heard more often than they are seen. They start to call soon after dusk, and end a series of croaks with a loud, distinctive scream.

The coloration of members of this genus serves as excellent camouflage among the epiphytes and other rain-forest plants. On all hyraxes, the dorsal area is a different color from the rest of the body. When threatened, *Dendrohyrax* turns this area toward its enemy. The hairs spread outward from a gland, leaving the naked center area exposed. The function of the gland itself, however, is not fully understood.

▲ *Golden cat* ▼ *Tree hyrax*

Duiker
Cephalophus sp.

African ungulates restricted to the rain forest are few indeed, especially when compared with the enormous numbers of species that inhabit the savanna. Most of those few that do make the forest their home—including about ten species of duikers, the royal antelope, and the chevrotain—share several basic characteristics. They are relatively small; they have excellent hearing; on most species the rudimentary horns are set far back on the head, thus minimizing the animal's chances of getting tangled in the undergrowth; and they are not gregarious. Their short legs are further indications of habitat—long legs would have little value in the closed-in areas of the forest, where running is difficult if not impossible.

Duiker is an Afrikaans word meaning "diving buck," and duikers are especially known for their habit of diving into cover when pursued. These shy and wary ungulates are meandering browsers on leaves and what fruit they can find on fallen logs or on vine-covered shrubby growth. They have also been known to eat insects, and in the few places where grass is available, they graze.

Bongo
Taurotragus euryceros

Stealthy, fleet movement through the forest characterizes this rather atypical rain-forest ungulate. The bongo is unusual because of its large size (up to 342 pounds [155 kg]) and its spiral horns, which are much longer than the headgear of most other forest ungulates. Bongos avoid getting the horns caught in brush by simply laying their heads back as they run; many bongos have bare spots on their backs where the horn tips touch.

These richly colored mammals are closely related to the kudu and the eland—so closely related, in fact, that many taxonomists have been somewhat uncertain about their classification. The bongo's most recent Latin designation places it in the same genus as the eland.

Bongos have few natural predators in their forest habitat. However, the setyot plant, a toxic plant that flowers only once every seven years, is thought to act as a check on bongo populations. Bongos eat the plant in great quantities, but it takes well over a year to affect the size of the herds. The giant forest hog may be subject to the same phenomenon.

Giant Forest Hog
Hylochoerus meinertzhageni

The wart hog is found in the savanna, the collared peccary in the desert, and the wild boar in the temperate forest. A fourth type of wild swine, the giant forest hog, inhabits the tropical rain forest. It is a large animal, up to 606 pounds (275 kg), and somewhat feared by the local people in its region because it is likely to charge without warning. This may simply be defensive behavior on the part of the males, which are highly protective of their five- to twenty-member social units, called sounders. In spite of the animals' ferocity, the skins of giant forest hogs often end up as the war shields of African tribesmen.

Unlike the wild boars, these hogs seldom feed by uprooting the soil. Instead they eat mainly leaves, berries, and fallen fruit, at least in the heart of the forest. Near the forest edge they feed on grasses and shrubs, and they can do a great deal of damage to domestic crops.

As conspicuous as these mammals are, they were not discovered until after the turn of the century. Their species name is derived from the name of their discoverer, Meinertzhagen.

Okapi
Okapia johnstoni

Okapis are the forest cousins of the giraffes. They live so deep in the forest that they were unknown even to the scientific community until the turn of the century, when their discovery surprised and excited mammalogists all over the world. These animals are wary indeed, and are adapted to sense danger much more acutely through their ears than through their eyes. They use their long tongues in typically giraffelike fashion—to strip young shoots from the branches of shrubs and trees. (The tongue is so long that the okapi bathes its eyes with it.) The okapi's real habitat is in forest clearings where tangled greenery has grown up, since food is unavailable where the canopy is completely closed overhead and sunlight is minimal. The diet of leaves is supplemented with some fruits and forest grasses; okapis also eat the charcoal of trees burned by lightning, as well as saline clay found near forest streams. Because of its unusual eating habits and its low reproductive rate (a single young is born after a fourteen-and-a-half-month gestation), the okapi is very hard to keep in captivity.

▲ Giant forest hogs ▼ Okapi

Moon Rat
Echinosorex gymnurus

Moon rats are among the larger insectivores. They resemble both tenrecs and hedgehogs, although they are really more ratlike than either of those two groups. Their most noticeable feature is an obnoxious odor, emitted from two glands in the anal region, that presumably serves some defensive purpose.

E. gymnurus, which is also known as a gymnure, has a very narrow body. Quite possibly this is an adaptation that makes foraging in small cracks and crevices possible. Moon rats are nocturnal, spending the day in burrows or other types of nests on the forest floor. By night they search for earthworms, aquatic foods, and insects. They swim well, and are properly furred for entering the water —their outer coat seems to protect the underfur from becoming saturated.

Tree Shrew
Tupaia sp.

Because tree shrews share a number of characteristics with certain primates—the bony rings that surround their eyes, for example, and their lemurlike grooming tools— taxonomists often classify them as lower primates. (The grooming tools are actually cartilaginous structures located under the tongue. They clean the fur that the animal threads through them, and they also serve as a pick for the lower front teeth.) Some taxonomists, however, group the tree shrews with the insectivores. Insects do make up a large part of their omnivorous diet, itself a major factor in the tree shrew's continued survival.

Tree shrews are extremely un-primatelike in their manner of reproduction and caring for the young. A female usually bears two offspring per litter, in a separate nest from the parents' sleeping quarters. After being cleaned and nursed, the newborns are virtually abandoned except for ten-minute feeding sessions every forty-eight hours.

The combination in one animal of all these characteristics, plus the fact the the tree shrew is very squirrellike in appearance, results in the species' being regarded as something of a transitional animal between two orders. The tree shrew and elephant shrew, in fact, have been assigned their own order, Menotyphla, by some scientists.

Colugo, or Flying Lemur
Cynocephalus variegatus

Flying lemurs are superb gliders, able to cover up to 443 feet (135 m) in a single glide without losing more than 40 feet (12 m) or so in elevation. Their furred membrane is the key to such movement. It extends all the way from the tip of the tail to the neck, and covers the animal's webbed feet. When open, it creates a parachutelike gliding structure that is more efficient at harvesting the wind than the gliding membranes of either the flying squirrels or the members of the Phalangeridae family. At rest, these mammals hang from a branch by their claws, with their head held up.

Like tree shrews, colugos have puzzled taxonomists. They have lemurlike grooming incisors in their lower jaws, yet possess very primitive skulls much like the insectivores'. Today they are classified in their own order, Dermoptera. The name is taken from two Greek words, one meaning "skin" (*derma*) and the other meaning "wing" (*pteron*). The order has only one genus and two species.

Rousette Fruit Bat
Rousettus sp.

Fruit bats of the genus *Rousettus* have large eyes indicative of their good vision. Yet they also use echolocation to navigate dark caves, which are their most common homes. The sound pulses they emit are unusual in that they are produced by the tongue rather than by the larynx. The individual pulses are short and of a low frequency with no discernible pattern. They are loud enough to be heard by human beings standing several feet away.

Some *Rousettus* species form maternity colonies much like a hospital's maternity ward, rare among the fruit bats. The males, which may also form their own groupings, cradle their young in their wings whenever the females will permit them to come near. Infighting dominates the colonies of at least some species, to the extent that large groups of individuals may leave their caves in great sweeping flight, returning only when the disturbances are settled.

▲ *Rousette fruit bat* ▼ *Tree shrew*

▲ *Slow loris*   ▲ *Male proboscis monkey* ▼ *Mindanao tarsier*

Slow Loris
Nycticebus coucang

Because of its very deliberate hand-over-hand movements, this prosimian was at one time mistaken for a sloth. The two animals do occupy the same arboreal niche; the loris creeps along on the upper and under sides of branches as it forages for insects, small vertebrates, fruits, and greenery. The loris's muscles are specially equipped for this life style, and they do not suffer from the fatigue that affects other animals' muscles after prolonged gripping. This is because blood flows freely through them, even after they have been in the same position for several hours. In addition, the loris's thumb and big toe are opposable, enabling the animal to grasp easily and to maintain its lock on branches.

Lorises spend the day asleep, rolled into a ball. The curled-up position probably serves to keep in body heat, since the loris has no tail with which to insulate itself. And although it inhabits the tropics, it has an extremely low metabolic rate and would probably freeze to death were its coat not as warm as it is. Rolling into a ball also plays a part in self-defense, since lorises sometimes elude predators by assuming the curled-up position and then dropping from a limb.

Mindanao Tarsier
Tarsius syrichta

Tarsiers are as quick in their movements as slow lorises are deliberate in theirs. In fact, they are the most highly specialized leapers among the prosimians. The long tarsal bones in their feet, from which their name is taken, provide the leverage to leap up to 5 feet (1.5 m) at a time. Tarsiers are clingers, their fingers and toes terminating in powerful suction pads. When used in conjunction with their supporting tail, these pads enable the animal to maintain a completely vertical position on a tree trunk, even when fast asleep.

The structure of the tarsier's face also helps the animal to execute its leaps. The head itself rotates almost a full 360 degrees, which affords a wide field of vision. And the huge eyes, set close together at the front of the face on either side of a small nose, provide the depth perception necessary for judging distances accurately and for locating prey by sight. The tarsier's greater reliance on vision than on smell—a characteristic of the true monkeys and apes—represents an important evolutionary advance, and has helped taxonomists determine how the tarsier fits into primate evolution: it is considered the most advanced of the lower primates.

Proboscis Monkey
Nasalis larvatus

One of the most striking examples of sexual dimorphism in the animal kingdom is the difference in nose size between male and female proboscis monkeys. Sexually mature males (those over seven years of age) have fleshy noses so enormous that they may occasionally get in the way when the monkeys eat. The nose stiffens each time the animal produces its peculiar honking noise, which is used to warn troupe mates of nearby danger. The female has an upturned and much smaller nose, and its proportions do not change much as she matures. She makes a sound much like a goose.

Proboscis monkeys are leaf-eaters. They are extremely fond of fresh water, and will take advantage of nearly any opportunity to paddle about, dive, or simply lounge in its vicinity. For reasons that are not totally understood, these monkeys are difficult to keep in captivity.

Crab-eating Macaque
Macaca fascicularis

Macaques are a varied group of Old World primates characterized by a rather generalized diet, a terrestrial life style, and the presence of cheek pouches for transporting food. The genus includes the rhesus monkeys, from which the Rh factor in blood was isolated, and the Barbary apes of Gibraltar, which probably number fewer than twenty-five today.

Only slightly less famous than these two relatives is the crab-eating macaque. As its name implies, it frequents not the forest proper, but mangrove swamps and similar areas where crabs and other crustaceans are plentiful. *M. fascicularis* crosses bodies of water readily to reach islands, and the fact that it has become adapted to regions that have undergone some deforestation makes the species' continued survival look hopeful.

Just how and why these macaques came to occupy their rather unusual niche is a matter of speculation. It may be that the more typical primate niches had already been filled by the time these monkeys evolved. To survive, therefore, they had to exploit the food of the coastal regions, the only habitat still open to them. Since the only substantial food in that habitat is crustaceans, that became the staple dietary item.

* Orang-utan
Pongo pygmaeus

Female orang-utans weigh about 88 pounds (40 kg), males about twice that. Yet they transport themselves through the treetops of Borneo and Sumatra by their toes and hooked fingers, and often travel as much as 60 feet (18 m) aboveground. To be sure, their movements are deliberate—they grasp a branch with one limb before releasing their hold with the other three. And they travel on the middle part of the branches rather than on the less stable ends. Nevertheless, these largest of arboreal animals present quite a spectacular sight in the trees. They construct a new tree-nest for sleeping each night, and take shelter under branches that they hold above them like umbrellas during daytime rainstorms. On the ground, they walk on the outside edges of their feet.

Orang-utans are solitary, an unusual characteristic among the primates. Rarely do they indulge in mutual grooming, which the more social chimpanzees, for example, do frequently. Their throat pouches, much more pronounced in males than in females, are thought to act as resonators for the sounds the adult males make, long calls that can be heard up to a mile (1.6 km) away. These help the males to know each other's whereabouts, and thus to stay apart.

Females breed infrequently, usually every three to four years. Because of irresponsible capturing and handling of orang-utans for zoos, and because of timbering operations that have destroyed much of their habitat, they are thought to number only about five thousand individuals today.

Lar Gibbon
Hylobates lar

Among the apes, gibbons are the undisputed champions of brachiation. They have very long forearms, and they propel their bodies hand-over-hand among the branches so swiftly that they actually seem to be airborne at times. This primate executes midair turns with ease, and often covers as much as 10 feet (3 m) in a single swing. The gibbon is also remarkably agile and precise in its movements—in midair, it can pluck food from a tree with one hand while the other hand grasps for a branch to swing to. On the ground gibbons both walk and run upright, holding their long arms out for balance. It is this freedom of movement in their forearms, in fact, that sets these animals (and the orang-utans, chimpanzees, and gorillas) apart from the monkeys.

It is interesting that these fast-moving animals are very particular eaters, carefully removing the skins and even small imperfections from fruits before they eat them. Their long fingers allow for rather delicate manipulation.

Prevost's Squirrel
Callosciurus prevosti

The Prevost's squirrel is one of a large group of Asian rodents known as the beautiful squirrels, which is what *Callosciurus* means in Greek. All the members of this genus behave in rather typical squirrellike fashion, but are atypical among rodents in their coloration. While a few have somewhat mottled coats, many are strikingly solid white, olive, cinnamon, or bright chestnut, and others have distinctive two- or three-part colorations. The fur is not unusually soft, but it is full and rich-looking, even through the tail. Beautiful squirrels are sold as good-luck pets in Formosa, where they are kept in hanging cages.

In the wild these rodents live primarily in the trees, where by day they forage for fruits, nuts, and flowers, and, very likely, for the eggs of birds and insects. They make their homes in tree hollows as well as in the more typical leaf-and-stick nests. The length of gestation is unknown. The typical litter consists of three or four young, although females have up to six nipples.

Brush-tailed Porcupine
Atherurus sp.

In addition to the sharp body spines characteristic of all porcupines, brush-tailed porcupines and the better-known crested porcupines possess rattle quills as protective devices. The brush-tailed species' rattle quills are long and slender, with every other segment flattened in the same plane; the crested species have gobletlike rattles. The sound the animals make by rattling these quills is very similar to that made by rattlesnakes, and sometimes has the same effect, that of deterring enemies.

Brush-tailed porcupines swim well and spend most of their time on the ground rather than in trees. They are totally nocturnal; during the day they take shelter, groups of up to eight individuals sometimes burrowing together. Usually they make their burrows among tree roots or under rocks and in stream banks. They forage in groups for various vegetable foods and insects. During cold weather, brush-tailed porcupines reportedly go into a state of dormancy

Indian Dhole, or Red Dog
Cuon alpinus

Much has been written concerning this carnivore's ferocity. Dholes hunt in packs of five to forty members, and the larger packs can bring down leopards, tigers, bears, and other large meat-eaters. Typical quarry, however, are various deer, antelopes, and other wild ungulates. The dhole often chases these for some distance, relying on its endurance rather than its speed in pursuing prey. It often bites and rips the flesh off its victim's flanks while both pursued and pursuer are still in motion. If there is enough space, a pack of dholes may confuse their prey by whizzing around and around it, making only two or three dholes seem like a dozen, at least to the victim.

Dholes are not true dogs, since they have fewer teeth and more teats than dogs do. (Though both dogs and dholes belong to the same family, Canidae, dogs are classified in the genus *Canis*.) They also howl instead of bark, and have slightly different muzzle profiles from dogs. Although basically forest animals, they can tolerate a greater range of climatic conditions than any other wild canid or similar form.

Malayan Sun Bear
Helarctos malayanus

This smallest of all bears, only 55 to 143 pounds (25–65 kg), is an excellent tree-climber, having hairless soles on its feet instead of the hairy soles typical of ground-dwelling bears. By day it nests and sunbathes in a crudely formed structure of branches it has built some 10 to 23 feet (3–7 m) off the ground. By night it searches for fruits, bees' nests, and insect larvae. Termites are also a favorite fare, and the bear will thrust a forepaw into a termite colony, then lick the insects off its paw with its elongated tongue. The long claws on the forefeet are used, in fact, in much feeding activity, tearing open the nests of bees and ripping the soft part of coconut palms into manageable pieces, for example.

Although adult sun bears can be quite dangerous, the cubs and juveniles make marvelous pets. Reportedly, they adopt particular mannerisms such as sucking a paw and making a noise rather like a contented hum before falling asleep. And accounts of their various antics indicate considerable intelligence on their part.

Binturong
Arctictis binturong

The binturong is one of the most oddly assembled mammals in the world. It has tufted ears and a shaggy coat that, like the coat of the sloth, often takes on a greenish color, probably from the algae living in it. This feature provides marvelous camouflage in the treetops. Unlike the other civets, to which this mammal is related, the binturong walks plantigrade like a bear; in fact, it has even been said to resemble the bears, particularly since it measures up to 6 feet (1.8 m) in length. Most strangely, perhaps, the binturong is a carnivore whose diet consists primarily of fruits. Its teeth are blunt and relatively small.

Except for the kinkajou, this is the only carnivore with a prehensile tail. As it climbs in the treetops, it uses the tail to steady itself, usually uncurling it slowly from one branch, then wrapping it around another. Juveniles may hang by the tail, but adults probably weigh too much to be able to do this.

* Bengal Tiger
Leo tigris tigris

The regal tiger seems doomed to extinction, despite valiant preservation efforts on the part of international conservation organizations. Its prey species, such as wild boar and deer, have already overpopulated areas where the tiger has been wiped out, and they now pose serious threats to agriculture. Yet poaching for tiger skins continues, as do the trapping of tigers for the protection of domestic species and the wholesale destruction of forest habitat.

Tigers are not fast runners, but hunt stealthily under cover until very near their prey. (Their coat provides wonderful camouflage in the forest.) Then they rush the prey, and strangle it as soon as they bring it to the ground. Sharp retractable claws, a formidable set of teeth, strong forelimbs, and massive shoulder muscles are the tiger's greatest weapons. Tigers eat up to one-fifth of their weight at a single kill. If they fail to finish a carcass, they may bury it and return to it later. Hunting skills are acquired by young tigers only after many trial-and-error sessions with their mothers. These and other family activities are of vital importance, since tigers (unlike lions, for example) are solitary and do not have a social structure such as a pride that allows for daily contact with "aunts" and several other individuals.

Water is essential to tigers, for bathing and drinking. In certain places where fresh water is scarce, tigers have become adapted to drink saline water, which is thought to damage the liver and kidneys. It may be that these physiological changes lead to increased aggressiveness toward people and help to account for the man-eating habits of tigers in certain places.

Muntjac
Muntiacus sp.

At 500 to 700 pounds (227–318 kg), sambars are the largest deer of the Asian forests; muntjacs are the smallest, weighing only 31 to 40 pounds (14–18 kg). Known also as barking deer, muntjacs produce a barklike sound when alarmed or aroused. If a tiger or some other enemy is nearby, the muntjac may continue its warning bark for more than an hour at a time. Barking is unusual in a solitary species such as this one, and it serves to alert all the nearby prey species.

Only the male muntjacs have antlers, which grow from bony pedicles that extend well down the face. (Another alternate name for the muntjac—rib-faced deer—comes from these facial ridges.) The antlers usually measure less than 6 inches (15 cm) in length, and seem considerably less useful as a defense weapon than the males' upper tusks. Actually modified canines, the tusks are quite capable of injuring assailants. Females have smaller tusks than the males.

The muntjac's breeding season falls near the end of summer, which is relatively early for a deer. Although the gestation period is known to be six months, the exact process of securing mates is unknown. It may be that, like many other ungulates, muntjacs assemble harems.

▲ Bengal tiger ▼ Gaur

Gaur
Bos gaurus

Wild cattle seem to be of extreme sizes in Asian tropical forests. The anoa and the tamarau are among the world's smallest, and the gaurs are extremely large. Males weigh up to 2,200 pounds (1,000 kg), the females about three-quarters of that. Herds are of mixed sexes, although old bulls tend to be solitary and young ones to form separate groups. The herds are considerably smaller than buffalo herds, and rarely contain more than fifteen to twenty animals. They spend most of the daylight hours chewing the cud in the cooler parts of the forest, then go out into clearings to feed at dawn and dusk.

In spite of their size and numbers, gaurs are shy and easily intimidated. Although they retreat from danger if they can, they have a unique form of self-defense (or possibly of display only, as some authorities think): bulls approach their opponents broadside instead of head-on. If they strike with the horns, they do so with the sides of the horns. The height of the animal at the withers is no doubt a threat, especially viewed from the side.

Reproduction is geared to periods of heaviest rainfall, with the result that fresh grasses are usually available to new mothers and their offspring. This timing is important, some periods being more conducive to regeneration than others, even in the tropics.

Duck-billed Platypus
Ornithorhynchus anatinus

The duck-billed platypus probes with its leathery bill along stream and lake bottoms, searching for mollusks, crustaceans, and other aquatic foods. As the platypus submerges, its ears and eyes close, making the animal dependent on its refined sense of touch. A flattened tail and webbed feet are helpful in swimming. When the platypus leaves the water, part of the webbing gets folded into the palm, leaving the nails exposed for burrowing.

Platypuses excavate two types of burrows. One is used by the male alone during the breeding season. The other, a nesting burrow, is made and used by the female alone, and is dug into a bank with its entrance placed far above water level. After mating, the female isolates herself in the burrow, plugging the entrance behind her. About two weeks later she lays one to three soft-shelled eggs, and incubates them for about ten days. Then they hatch. Like the young of echidnas (the only other representatives of the monotreme order), platypus young nurse not from nipples but directly from milk that oozes onto the mother's abdomen. This characteristic, and an inconsistent body temperature (it generally ranges between 72 and 95° F [22–35° C]) serve most dramatically to set the monotremes apart from the more highly evolved mammal forms—the marsupials and placentals.

Striped Possum
Dactylopsila sp.

Striped possums feed in much the same manner as the aye-aye, by ripping open the bark of trees with their incisors, then extracting insects from the tree's crevices with their long, skinny fourth finger, which is adapted for just that purpose. Tapping the bark and then listening and smelling for their insect food is a constant part of their activity. These possums seem to have no preference about feeding position; apparently they are just as comfortable upside-down as they are right-side-up.

Although striped possums bear markings similar to a skunk's, they do not eject a foul-smelling liquid as skunks do. They do, however, have glands that exude a repugnant odor, which causes their bodies to smell. This odor repels at least some potential enemies that might otherwise disturb the nesting quarters in hollow trees.

Green Ringtail
Pseudocheirus archeri

Green ringtails are most unusual in their coloration. They have two stripes down their backs, and their fur has a greenish-golden tone. The color seems to result from a particular combination of yellow, black, and white pigments that, because of the animal's body structure, blend together in a distinctive way. This mammal looks like a ball of green fluff as it sleeps by day curled up in a vine or on a branch in the rain forest.

Like all members of this genus, the green species has a very prehensile tail, with which it constantly grips and climbs. When not in actual use, the tail is usually carried in a tightly coiled position. Fruits, flowers, leaves, insects, and small vertebrates make up this possum's diet.

These mammals, like cuscuses, are classified as members of the large and somewhat confusing family of marsupials known as the Phalangeridae. The members of this group range from the tiny pygmy and honey possums to much larger species, and also include very specialized gliding mammals such as the feather-tail glider.

Tasmanian Devil
Sarcophilus harrisii

Tasmanian devils are marsupial carnivores, and along with the native cats and several other animals, occupy a carnivorous niche in the Australian realm. Dingoes, seals, and bats, in fact, are the only nonmarsupial carnivores in the entire region.

This is a rather feisty mammal, and its ears actually turn red whenever it is enraged. It snarls, growls, coughs, and even barks, and it has the equipment to back up these threats—its teeth crush bone quite easily, and its digestive system can handle the feathers, the fur, and the bones of prey animals. The Tasmanian devil takes a wide variety of creatures, ranging from small mammals and birds to species considerably larger than itself. The name "devil," given it by the white settlers of Tasmania, certainly seems appropriate for a mammal such as this.

Spotted Cuscus
Phalanger maculatus

The Phalangeridae family is a large and diverse group of marsupials. Some of its members resemble mice, others resemble prosimians, still others resemble squirrels, and the koala has no really close look-alikes. Australia, of course, has no native squirrels or prosimians, and members of this family fill ecological niches occupied by those types of animals in other regions. They are arboreal and for the most part herbivorous.

The spotted cuscus has opposable digits well suited for grasping, and it usually carries its prehensile tail rolled up tight. Scales cover the tail's lower half and help the cuscus to grip tree limbs. The animal's movements are extremely slow, yet serve it well since arboreal predators are almost totally lacking, and there is little if any need for speed. Many human beings, however, traditionally relish the flesh of cuscuses and eat these mammals whenever they can.

Tree Kangaroo
Dendrolagus sp.

As they evolved, tree kangaroos moved from the trees onto the ground along with their terrestrial relatives; then, much later, they moved back into the trees. Although adaptations never really reverse themselves, today's tree kangaroo is much more obviously an arboreal animal than its cousins on the plains. It has limbs of nearly equal proportions, no longer needing strong hindquarters for hopping. And its hind feet are equipped with rough soles that help the tree kangaroo grip the branches and keep it from slipping. The long, curved foreclaws serve the same purpose. The noticeably long tail serves both as a rudder when the animal is leaping about on the ground and as a brace when it is climbing. Tree kangaroos are not particularly good climbers, yet manage well enough since they have no arboreal predators. They sleep and eat in the trees, feeding primarily on vegetation that they pluck from the branches.

Koala
Phascolarctos cinereus

Several species of possum, the tuan, the numbat, the koala, and some gliders all inhabit a type of forest known as the eucalypt forest. It is not a true rain forest, nor is it very similar to the vast temperate-forest belt that covers much of the northern hemisphere. Many of its trees are very tall, and the older ones have hollows in them. These make wonderful homes for marsupials.

Koalas are by far the most well-known eucalypt dwellers. Their diet consists almost exclusively of the leaves of a few eucalyptus species; their digestive tract is specially adapted to handle the bulky greenery. (A single meal might consist of 2½ pounds [1 kg] of leaves.) The koala is equipped to tear the leaves from their branches with its very secure grasp (the first two and last three digits work in concert to provide gripping power). Long forearms that end in sharp claws are a major aid in climbing smooth-barked trees.

The single offspring of the koala remains in its mother's pouch for about the first six months of its life. After that, the mother weans it by letting it lap a special substance of predigested eucalyptus leaves from her anus. This serves as a transitional food between milk and foliage. For the next six months or so, the mother carries her offspring piggy-back or hugs it close to her body.

Sugar Glider
Petaurus breviceps

Flying mammals are represented in Australia by the gliders (also called gliding possums), among others. The main difference between these marsupials and flying squirrels is their fully furred tails, as opposed to the flying squirrels' more flattened ones. Tails are used not only to help the gliding animal steer, but also to collect nesting materials. The flying phalanger hangs from a branch by its hind feet while it gathers leaves with its forefeet. It then passes the branches from the forefeet to the hind feet to the tail, which curls around and holds them while the animal runs to the tree hollow that will become its nesting site. Nests often hold more than one generation of animals at a time, and are heavily marked with urine.

The exact distance one of these mammals can glide is a matter of some debate. A claim has been made for 164 feet (50 m), but this may be accurate only when the starting point of the glide is relatively high above the ground. In other words, the animals lose considerable altitude during leaps. They land head-down, and jump into the air to turn themselves head-up before proceeding to spiral their way up a tree trunk.

Numbat, or Marsupial Anteater
Myrmecobius fasciatus

Numbats provide one more example of marsupials filling niches in Australia that are occupied by other types of mammals on less isolated continents. In this case the niche is that of a small carnivore, most specifically that of an anteater. Numbats have stout claws that can rip open logs suspected of housing termites. And their 4-inch-long (10-cm), agile tongues flick through the termites' nesting chambers with incredible speed, picking up the insects in the process. After a filling meal, the numbat may stretch itself out on a log with its tongue dangling.

This lovely, dramatically colored mammal is unusual in that it moves about by day instead of in the evening. It is not unlike a pointed-faced squirrel in general appearance. Eucalypt woodlands are the numbat's habitat, and tree hollows are its favorite sleeping places. Habitat destruction and predation by dogs, cats, and foxes have caused reductions in this mammal's numbers.

Red Flying Fox
Pteropus scapulatus

Flying foxes, so called because of their caninelike heads and faces, are among the largest bats alive. A wingspread of 5 feet (1.5 m) is not unusual, nor is a weight of some 28 to 31 ounces (800–900 g).

Pteropus bats traditionally eat fruit, and are known in Australia for raiding orchards. It is thought by some people that they may turn to cultivated fruits when natural foods are hard to obtain, though this is still debatable. Their tongues are affixed to the floors of their mouths, and cannot be extended very far. They are adapted to handle fruits and blossoms, the bats' simple digestive systems processing mainly the juices of fruits crushed by the molars. Hard material is generally not digested. However, the fruit bats do pass seeds and berries through their digestive tracts, and so inadvertently help in the dispersal of tree seeds.

Flying foxes are excellent climbers, their forelimbs being equipped with two claws instead of the more usual one. They live in large camps, or aggregations, where they may be heard "chattering" and moving about quite noisily.

2

The Temperate Deciduous Forest

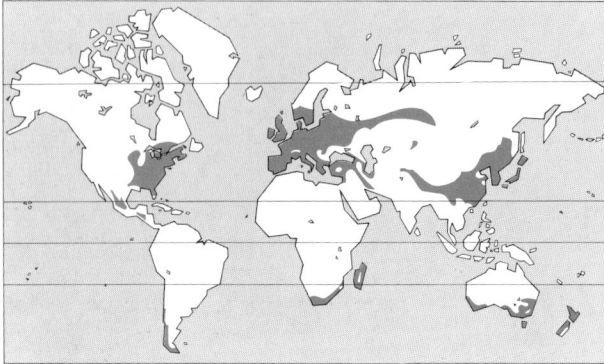

Temperate deciduous forests are composed principally of trees that shed their leaves each autumn. These forests once stretched over most of central and western Europe; over eastern North America; and, in East Asia, over much of northern and central China, Korea, and Japan. Although these regions are now heavily developed (particularly in Asia), some stands of original forest still remain, stands such as the United States' Great Smoky Mountains National Park. They provide a glimpse of this complex biome in its untouched state. (Southern Chile and a small part of the Australian realm have deciduous forests also, but they are tiny compared with the mid-latitude woodlands of the northern hemisphere.)

Rainfall in the deciduous forest is both plentiful (27–58.5 inches [70–150 cm] per year) and well distributed throughout the four seasons, which are clearly differentiated. In the more northerly regions snow may fall several times during a winter. Yet the annual growing season is relatively long, 250 days or more in the southern parts of the biome. The extensive periods without frost allow each generation of plants ample time to undergo the cycle of leafing, flowering, and bearing fruit and seed. This is a luxury that many plants of less temperate climates cannot afford.

Soil, too, is of a type that encourages lush plant growth. Known as mull, it is composed of decayed leaf litter mixed with mineral matter. The fallen leaves decay rapidly due to the humid environment and to countless ground organisms such as earthworms and fungi. Earthworms actually eat their way through the ground, and as they do so, they churn and mix and digest and excrete and reassemble the ground particles. Thus they help create an extremely fertile soil very quickly.

It is the trees, of course, that give this biome its distinctive character. Their broad leaf surfaces are indicative of the enormous amounts of moisture they transpire, moisture that is usually felt as a refreshing humidity by human beings as well as by other types of mammals.

Deciduous, or broadleaf, trees change their dress with the seasons, and are perhaps best known for their brilliantly colored autumn leaves. The autumn color change takes place in the following manner. The cool, dry days of early fall trigger the formation of a layer of tree cells known as the abscission layer. Located just at the base of each leaf's stem, this layer and another one under it seal off the dark green leaves of summer, preventing them from obtaining the nutrients and water necessary to photosynthesis. Without these nutrients, of course, food production ceases and the chlorophyll disappears from the leaves. Then the lovely red and yellow pigments, which are actually present at all times, are unmasked.

This process (and the leaf-fall that follows it) is a tree's way of protecting itself against winter drought. Tree roots are able to absorb less and less moisture as soil temperatures move toward freezing. If a tree retained its leaves, frosts could cause it to experience severe drought. By losing the leaves, however, a tree halts its food production and transpiration processes, and thus conserves its moisture.

While denuded winter branches have their own kind of beauty, they can be stark and forbidding. They also severely limit the number of sheltering sites available to wildlife. But they allow a relatively great amount of light to strike the forest floor. In response to this abundant light, many low-growing woodland species leaf and flower very early in spring, before the trees produce foliage that blocks the sun. Known collectively as the herb

Preceding pages: In winter, with most trees bare, white-tailed bucks move through a Louisiana forest. Buck on right has tail raised to signal danger.

layer, these plants include many, many types of delicate and lovely wildflowers, as well as ferns, mosses, fungi, and other organisms without woody parts. Perhaps this is the loveliest moment in the forest, when wildflowers such as hepatica and trillium spread in every direction and the phacelia grows so thick and lush that walking through its carpet of blossoms is unthinkable. As the growing season progresses, the forest floor becomes gradually shaded by tree leaves. Eventually, shafts of sunlight dance across it only where the trees are thinnest, and by late summer most of the fragile blooms have vanished under the thick canopy of mature greenery.

Above the herb layer grows a layer of plants known collectively as the shrubs. Most of them are between 3 and 20 feet (1–6 m) in height, have at least two stems, and, like the trees, are broadleaf plants. Shrubby species vary considerably from one locale to another, and often grow only in association with specific other plants. Like the wildflowers and the trees, forest shrubs are marvelously lush and varied. Some of them, such as the sweet shrub, bear intensely fragrant blossoms.

Understory trees overhang the shrub layer. Usually no taller than about 33 feet (10 m), they may be the young or crowded individuals of tall tree species, and because of the crowding may not survive to maturity. Or they may be altogether different kinds of trees that are normally relatively short. (Dogwoods are a classic example of the latter.) The understory and shrub layers are the parts that give the deciduous forest its dense appearance, the tallest trees and the herb layer being too high and too low respectively to attract first notice.

The dominant trees in a deciduous forest may reach a height of 200 feet (61 m) or more. They are known collectively as the canopy layer, since their leaves form an almost lightproof cover above the other broadleaf plants. The trunks of deciduous trees are almost as important to the intricate workings of this biome as the leaves are, since they provide numerous animal homes and transport materials necessary for photosynthesis.

The deciduous forest is, in a very real sense, a mis-nomer. There are actually several types of deciduous forest (or more properly, of forest associations), each of them named for its dominant species. Typical associations are the oak-hickory forest, the magnolia-oak forest, and the beech-maple forest. The pine forests of the southeastern United States are also considered part of the temperate deciduous belt, even though the trees are evergreens. And for present purposes, the Mediterranean type of vegetation is treated as part of this biome.

What is called the mixed deciduous forest has more than one dominant species. For example, the tulip tree, the basswood, the buckeye, and the beech may all be found growing very near each other in certain locations. Where each type of forest association occurs, of course, depends on a combination of environmental factors such as soil, latitude, and rainfall.

In the case of deciduous-forest mammals, "forest-dwelling" does not necessarily imply "exclusively arboreal." In fact, many temperate-zone species, including the raccoon, the opossum, the eastern chipmunk, and even the so-called tree squirrel, are at least part-time ground-dwellers. This is hardly surprising, since by definition deciduous trees lose their leaves in winter, and hence offer little in the way of shelter or food to attract mammals.

"Forest-dwelling" does, however, imply "adaptable" in a very special sense. While the broadleaf woodland itself does not have the richly diverse growth of a tropical forest, it does offer its mammals many, many choices in terms of vertical living space (the forest layers). It also has pronounced seasons, which the tropical forest does not, so mammals must be able to cope with seasonal temperature changes if they are to survive. No other biome offers this combination of choices to its inhabitants.

The ways in which temperate-forest mammals respond to this challenge are nearly as varied as the species themselves. Animals such as the muskrat and the Chinese water deer spend much of their lives in or near the water. Water, a stabilizer of temperatures, is usually plentiful in this biome. The streams and rivers seldom run dry, and in fact far more animals are probably destroyed through flooding than are lost through real drought.

Other types of animals live almost exclusively underground. Moles and shrews burrow, for example, not merely to escape the heat or cold, as many desert and tundra mammals do, but also to take advantage of the banquet of earthworms and other tiny organisms that the humus offers.

Still others have taken advantage of man's penchant for clearing trees, and have come to inhabit what is known as the forest edge. The forest edge may occur naturally where biomes intermesh or even where bodies of water interrupt the woods. Or it may result from man-made clearings that have been created nearby.

Because light is plentiful in the forest edge, the shrubby undergrowth is usually greater here than in the heart of the forest. This means that there is ample shelter and there are more plentiful supplies of food such as berries and ripe young buds. Deer browse on the upper parts of many forest-edge plants, while rabbits, hedgehogs, and other species feed from the lower layers.

Of course deciduous-forest mammals are not without structural adaptations that help them to live efficiently. Many of them have well-developed auditory and olfactory senses, and communicate through them much more than through sight. This is only logical, since seeing over great distances is nearly impossible where trees and shrubs grow in almost every direction.

Many nocturnal mammals have evolved a special layer behind the retina of the eye, which helps them see in the dark. Called the tapetum, this layer reflects light in the same way that roadway reflectors do. (Incidentally, this layer is what causes mammals' eyes to glow when strong light strikes them.) In the broadleaf forest, where light is often minimal, this adaptation is a precious one.

The opossum and the white-footed mouse are both endowed with prehensile, or grasping, tails. Certainly they contribute to the animals' climbing prowess. And the gray fox is an excellent tree climber, a very unusual forest niche for a carnivore. Certain dormice are adapted to an

Autumn foliage in Michigan forest

almost exclusively arboreal life style, and no mammal is more at home in the trees than the flying squirrel, its specialized gliding locomotion made possible by a collapsible membrane. Bats, of course, fly quite as easily as birds.

Woodchucks find even the temperate-zone winters too severe to remain active, and bats, being basically tropical animals, generally cannot withstand prolonged cold weather. Therefore both hibernate, and some bats migrate. Many other types of mammals,however, seek protected homes during inclement weather without actually undergoing a true hibernation. When a warming spell occurs, even in midwinter, they may be seen wandering about in search of food.

Breeding season is rather uniform among deciduous-forest mammals. With the exception of deer, bats, and several others that breed in fall or winter, mammals tend to mate in late spring and early summer.

Perhaps the most important general statement that one can make about the deciduous forest is that the same factors that have made it attractive to human beings for centuries also operate where other mammals are concerned. For the most part, this is not a hardship biome, but a rather inviting place to live, with its rich soil and consequent high vegetative yield, its humidity, and its equable temperatures. The irony of this situation, of course, is that as people have overtaken more of the biome for their own "civilized" use, they have eliminated many other large and therefore spatially demanding mammal species from its premises—the wolves, the bears, the cats, and other predators are almost gone. Conversely, it is no accident that rats, mice, and many other smaller species thrive where buildings and cultivated lands abound. Human beings have created favorable living situations for the one group just as they have devastated the populations of the other. (Nowhere has man's influence been more obvious than in Asia, where only a few mammals are left in the sparse patches of temperate forest.) Regardless of its losses, however, the broadleaf forest is a rich and enticing biome with a wonderfully varied array of mammals.

Female opossum with young ▲

Opossum
Didelphis virginiana

Over the course of its long stay on earth, this mammal has developed several specialized body parts, none more useful in the treetops than its prehensile tail. The tail is, of course, an invaluable aid in climbing, but it has other uses as well—the opossum may curl it around bundles of leaves that are transported to the animal's nesting place, for example. An opposable toe on the opossum's hind foot makes grasping from the rear possible, and it, too, helps in climbing.

This is the only marsupial genus that inhabits North America. Like other marsupials, opossums bear tiny and incompletely developed offspring after only a brief gestation period—in this case thirteen days. The tiny opossums continue to develop in the mother's pouch, into which they must first crawl, and where they stay to suckle for about two and a half months after birth.

Opossums defend themselves by "playing dead." This adaptation, plus the fact that they have more teeth (fifty altogether) than any other North American land mammal, helps to account for the enormous success of the species.

Lesser American Short-tailed Shrew
Cryptotis parva

Only a little more than 3 inches (80 mm) long, *Cryptotis* shrews are the shortest New World mammals; they are also the only insectivore genus represented in South America. (Members of the genus are also found in Central America, Mexico, and the eastern United States.) Like other shrews, they must eat voraciously and regularly just to survive. In temperate areas this is not the problem that it is in very cold or hot arid regions, where food is scarce or where body heat is really difficult to control.

Two or more *C. parva* individuals may cooperate in excavating tunnels: one will dig while the other disposes of loosened soil. When the tunnel is completed, the shrews may store their various types of food within. Individuals also share nests, which are made of shredded vegetation. In fact, gregariousness is a characteristic of this genus; other shrew genera, such as *Blarina,* remain strictly solitary.

Females may begin breeding when they are only three months old. The three to six newborns weigh about three-tenths of an ounce (1 g) each, and their skin is translucent. In two weeks they reach adult proportions.

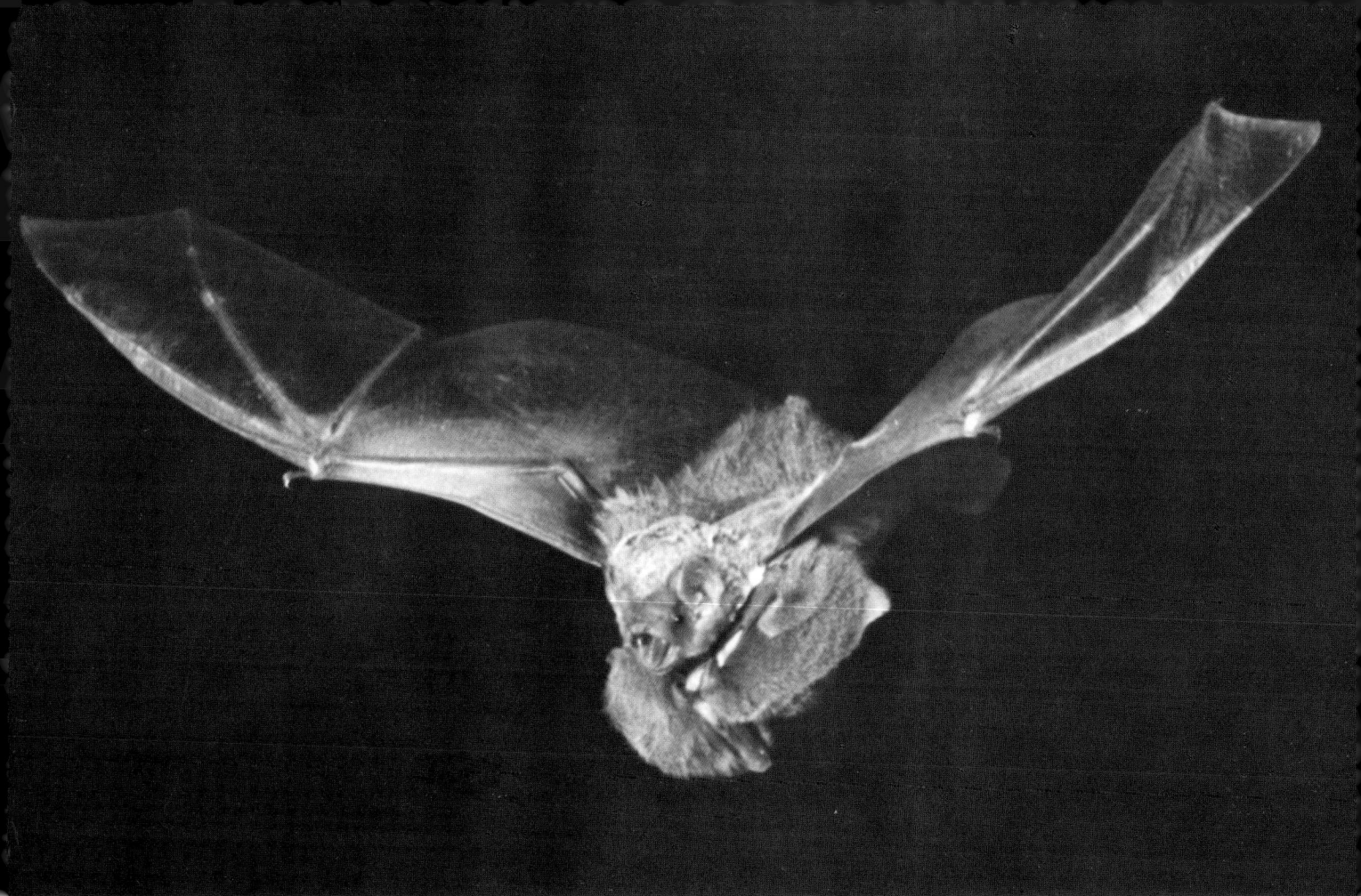

Female red bat in flight, carrying young ▲

Star-nosed Mole
Condylura cristata

The eastern half of temperate North America is the home of this mole with the bizarre nose apparatus. The twenty-two tentacles are its food antennae, always moving as the mole forages and contracted as it eats. What food the star-nose does not assimilate right away is stored as fat in the long tail. In late winter, when fresh food supplies are scarce, the mole is able to draw on this stored supply, which may also provide extra energy for breeding. It is not the only mammal to have such an appendage, some of the Australian marsupial mice being similarly adapted.

Thanks to waterproof fur, wide forefeet, and sturdy hind limbs, this mole is well adapted for swimming, and hunts underwater, sometimes even under the ice. It moves aboveground almost as frequently as it keeps to its subterranean rooms. In the mild winters of the deciduous forest it can even be found on top of the snow, instead of only beneath it, where most small tundra and taiga mammals stay for warmth.

Red Bat
Lasiurus borealis

Red bats are superbly adapted to the winter cold of North America's temperate zone. They have well-furred bodies, the greater part of which they cover with the tail membrane during hibernation. This allows them to adopt the nearly ball-like shape so efficient in cold-weather species. The red bat's metabolism adjusts rapidly to outside temperatures, and astoundingly, the species suffers no obvious ill effects if parts of its body completely freeze! It does not awaken from hibernation until temperatures rise above 66 to 68° F (19–20° C). Many other bats awaken when temperatures reach only 50° F (10° C) and so are forced to expend energy searching for winter food. By the time the red bat awakens, insects are already flying; thus it has little if any really difficult hunting to do. Because relatively high outside temperatures are required to awaken it, the red bat cannot hibernate in the fairly stable environment of caves; it seems to seek out trees in which to spend the colder part of the year.

As they fly, *Lasiurus borealis* females can carry a load of offspring that is almost twice the weight of their own bodies. The species travels fast, and is known to make extensive migrations, up to 1,000 miles (1,610 km).

Big Brown Bat
Eptesicus fuscus

Eptesicus fuscus has relatively little body fur or other insulation, yet it is one of the hardiest of bats, as evidenced by its ability to survive drafty hibernation sites in caves. During temperate-zone winters it is perhaps the most active of all chiropters. The adaptation that makes both hardiness and winter activity possible is a form of thermoregulation: these bats simply do not continue to hibernate under conditions of extreme cold. Rather, they awaken when temperatures fall much below freezing and seek out a warmer site. Obviously, for this species at least, the arousal and movement use up less energy than maintenance of body heat would.

During hibernation male *Eptesicus* bats tend to lose weight at a significantly faster rate than the females do. Explanations for the difference are little more than theory at this point. The phenomenon may be related to atmospheric conditions (including relative humidity) at the hibernation site, to how often the individuals waken and move, or to whether or not they cluster.

Experiments have been conducted on *Eptesicus* bats banded and released some distance from their homes. These studies show that *E. fuscus* has strong homing abilities—some individuals reached home from a distance of 450 miles (724.5 km) in a little over a month.

Keen's Bat
Myotis keenii

Myotis bats enter hibernation by allowing their body temperature to drop along with the outside air until the air temperature approaches freezing. Then the bat's metabolic rate increases enough to keep the body temperature above the freezing point. This type of temperature fluctuation has caused some observers to think that the bats' warm-bloodedness is poorly regulated, which is obviously not the case.

Several *Myotis* species are known to cluster during hibernation. Both the little brown bat, *M. lucifugus,* and Keen's bat cluster on walls in the warmer portions of caves. Clustering seems to insulate participants from various types of disturbances, including temperature fluctuations, and it may also fulfill some important social function. Those *M. keenii* individuals that do not cluster may winter under loose tree bark, or even behind window shutters.

Eastern Cottontail
Sylvilagus floridanus

This well-known lagomorph is among the most adaptable of mammals. Although it prefers to inhabit brushy areas along the forest edge, it also thrives in suburbia, particularly near orchards and crop lands. Cottontail homes are shared democratically, although some may be the exclusive property of particular individuals. In the coldest part of the winter, cottontails often seek shelter in woodchuck burrows, or in nearby woodlands or other protected spots. Almost any plant food will satisfy them, and they obtain moisture from dew, from snow, and from their food, as well as from freestanding water.

Large shoe-button eyes positioned on the side of the head allow rabbits to see in almost every direction at once. And the movable ears can be cocked to pick up the tiniest sounds, a distinct advantage in areas where rabbit hunting is popular. A rabbit's marvelous leaping ability is also a point in its favor.

Rabbit reproduction, of course, is legendary, especially the elaborate pursuits and games that precede mating. The tiny newborns are deposited in a well-lined underground nest that is covered with matted vegetation. The mother stays outside the nest, guarding her offspring. Periodically she returns to the nest to nurse her young. They are independent in only three to four weeks.

Woodchuck
Marmota monax

To a woodchuck, hibernation means both winter survival and oblivion. Even if a sleeping animal is handled frequently, its eyes remain closed; its temperature stabilizes at about 50° F (10° C); its breathing occurs only about once every six minutes; its heartbeat is slowed considerably; and its teeth and nails grow little if at all.

Not long after emergence, the males begin to wander from den to den in search of a mate. A pair remains together until near the end of the month-long gestation period. Then the male is banished from the den. Some males may find second mates, if there are females that have reached their fertile periods relatively late; the others spend the remainder of the spring alone. This adaptability on the male's part assures each fertile female of a litter, and the species, therefore, of survival.

These burrowing rodents (also called ground hogs) relocate about 400 pounds (182 kg) of soil each time they excavate a den. The aeration that such digging provides benefits the soil tremendously, and since a single woodchuck may use as many as four or five multichambered burrows at different periods, the species is extremely important ecologically. Summer dens are located in either grassy meadows or crop lands, winter dens in more protected spots in the forest or along the forest edge.

Woodland Jumping Mouse
Napaeozapus insignis

It is surprising to find a rodent built like this one in the forest. The tremendously long hind legs would suggest that the animal requires open ground for jumping, and this is true: these jumping mice inhabit only those regions of a woodland that have adequate open space to accommodate their method of locomotion. Areas near stream banks usually provide ideal habitat of this sort.

Napaeozapus climbs trailing vegetation almost as easily as it forages on the ground. Therefore it has access to a wide variety of foods, from vine seeds and berries to ground-hugging fungi and small invertebrates. Perhaps because of this, it stores no food, nor does it possess the cheek pouches characteristic of so many small rodents.

These jumping mice remain dormant from November to late April or early May. They mate upon emerging from their nests of vegetation, which are usually placed deep enough in the ground to protect the residents from frost. They may have both a spring and a fall litter before cold weather brings on dormancy once again.

Southern Flying Squirrel
Glaucomys volans

Although they are nocturnal and therefore not often seen by people, flying squirrels are important to forest balance. They rarely excavate their own dens, but nest in abandoned woodpecker holes. And they contribute to reforestation by burying various seeds and nuts. Their entire pattern of nighttime activity is very similar to the tree squirrels' daytime pattern, and in fact the two types of squirrels would probably compete were they not on complementary schedules.

Glaucomys moves its head from side to side before it launches itself in "flight." (It does not really fly, but glides by outstretching its gliding membrane, or patagium, which extends from the wrist to the ankle.) Perhaps it is taking a reading on the landing site, which is almost always the trunk of a tree. The glide itself ends in a parachute landing: the gliding membrane catches the air as the squirrel raises its tail and brings its limbs forward. Immediately, *Glaucomys* scurries to the opposite side of the tree trunk, to discourage predators. The entire range of aerial activities is carried out so swiftly and sure-footedly that casual observers may miss seeing the tiny gliding bodies altogether.

Eastern Chipmunk
Tamias striatus

Few other small rodents delight observers as much as this one does. Its coloration is rich and distinctive; its speed, small size, and inquisitive expression imbue it with a charming impishness; and its daytime habit of scurrying in and out among the stones of old walls and buildings near the forest edge makes this chipmunk seem almost like a clever wind-up toy. Yet it is an accomplished burrower, and in preparation for winter it stockpiles seeds in its complex underground chambers. Individuals may or may not go into a complete state of hibernation, depending on how far south they are. When the danger of frost is past, the chipmunk emerges to feast on a wide variety of animal and vegetable foods, usually while sitting on a favorite stump or fallen tree.

Eastern chipmunks are larger than the western and Siberian species. If they leave (or are removed from) the moist, shady atmosphere of the forest and venture into bright sunlight, they die quickly.

Rice Rat
Oryzomys palustris

Rice rats are just as at home in the salt marshes of the Gulf of Mexico or the Atlantic as they are in forest clearings or open meadows. Their only real habitat requirement is some sort of heavy ground cover in which they can forage without being obvious to their multitudinous enemies. Should they be pursued, they will dive and swim underwater for quite some distance to escape.

Rice rats may nest in the abandoned homes of marsh wrens, or they may weave their nests themselves. In either case, the nests are usually located well above flood level.

As its name implies, this rodent has a fondness for rice. Early colonists in North America were often perturbed to find these rats feasting on their freshly planted rice crops, as well as on the grain in its milky stage.

Cotton Rat
Sigmodon hispidus

The pine forests and grassy thickets of the southern United States are replete with this rodent. It is not entirely welcome, since it preys heavily on quails and their eggs, as well as on many valuable crop plants. Active throughout the day and night, an individual burns itself out and dies in a period of six to ten months, leaving a legacy of many offspring. The species' success is due in large part to two factors: the animals' nesting sites are always well concealed, and the young mature extremely quickly.

Networks of runways meander through the cotton rat's home range. These are the paths along which the animal moves and feeds, and they are usually marked by piles of clipped herbs and grasses. Harvest mice, rice rats, and shrews may use them also.

The species name, *hispidus,* comes from the Latin word for "rough," and refers to the texture of the animal's coat.

Muskrat
Ondatra zibethica

Marshes and rivers in various parts of North America and Europe are dotted with muskrat lodges. They resemble beaver homes in their overall shape, and are constructed of the same kinds of materials—sticks, cattails, and weeds cemented together with mud or peat. Sometimes a burrow in a bank is the preferred dwelling. In either case, muskrats enter their homes from underwater.

Swimming is second nature to these mammals. In fact, young muskrats can swim by the time they are only a week old, although their eyes do not even open until they are two weeks old. The vertically flattened tail provides sculling power and acts as a rudder when the muskrat swims. Movement in the water is accomplished with the weblike hind feet; the forefeet are held folded close to the body.

Muskrats are not rats at all, but are more closely allied with beavers. Their pelts are used by furriers; their sweet-smelling musk oil (secreted by the males to mark territories) is utilized by perfume manufacturers; and their meat is eaten by some people. Despite such high predation by man, this mammal is able to maintain its large numbers through a high reproductive rate.

Striped Skunk
Mephitis mephitis

Striped skunks are not true woodland creatures, but prefer the forest edge, meadows, and other areas clear of dense timber. For this reason they thrived as forests were felled during the opening up of North America's West, and they are still thriving over almost all of the continent except its northernmost regions.

This boldly marked species has no need to hide, so gifted is it at defending itself with its highly potent musk. The musk is discharged from tiny glands located beneath the skunk's tail, and its scent is so powerful that it is almost impossible to remove from clothing. The skunk is not trigger-happy with its spray, however. It must be seriously provoked or threatened before it takes aim, and even then it usually precedes the actual shower with some foot-stomping and a handstand or two.

Striped skunks resemble opossums in their habit of transporting leaves to a nesting site (skunks remain inactive in their dens in cold weather). Whereas the opossum uses its tail as a wheelbarrow, however, the striped skunk shuffles the leaves along beneath its body. Because they are omnivorous, striped skunks are able to forage successfully in areas not rich enough to support the larger carnivores or other species with very specialized diets.

River Otter
Lutra canadensis

River otters are the "playmate" mammals of North America. As far as anyone can tell, they simply enjoy the fun of a social outing, whether it be tobogganing, somersaulting underwater, playing with a rock, shooting the rapids, or an earnest game of tag. Their fun-loving nature is a testimony to their bright minds, since the few animal species that indulge in play regularly are also generally very intelligent. The river otter is no exception. Its curiosity compels it to examine nearly everything it handles, even if that means destroying objects in the process.

Lutra canadensis dives and swims beautifully, and is adept and graceful on land as well. Its broad webbed toes, so useful in swimming, also help the otter travel overland, especially as it alternately slides and runs in the snow. Eyes, ears, and nose are all designed to close tightly when the otter dives, and strong lungs allow it to stay submerged for as long as four minutes. This adaptation is especially helpful when the otter fishes, as it does fairly frequently. On land it may eat birds, insect larvae, and several other foods.

Gray Fox
Urocyon cinereoargenteus

Because trees are essential to the gray fox's existence, the range of this animal at any given time is determined to a great extent by their presence. In areas where large numbers of trees have been felled by man, the gray fox's numbers have declined drastically; and it is unable to colonize the areas of North America that are naturally treeless.

Their tree-climbing ability gives gray foxes an unusual position among wild canids. Since they have access to almost any kind of food, they have relatively little need to compete with many of the carnivores that are restricted to totally terrestrial niches.

Gray foxes climb by digging their forepaws into bark, gaining support from well-planted spikelike claws on the hind feet. Once they have ascended they can actually jump from branch to branch, to hide from enemies as well as to search for food. To reach the ground, they back down the trunks of trees.

▲ River otter ▼ Gray fox

Raccoon
Procyon lotor

Raccoons are a caricaturist's dream, with their well-defined facial masks, bushy ringed tails, and unusually sensitive paws. Their habit of "washing" foods when near water may be a form of sensual indulgence—the interplay of water flowing over held objects may simply feel good to the raccoon's hands.

These mammals arrive in the world with their climbing reflexes already developed. They can hug a tree trunk long before they can see it, though young raccoons may find themselves helpless when they try to descend. Although mother "coons" will direct their entire litters up a tree when danger threatens, experienced animals grow wary of this escape tactic, and rightfully so—many a raccoon treed by hunting dogs has ended up on someone's supper table!

An outside temperature of 26° F (-3° C) sends raccoons to their dens, which may be in hollow trees, caves, or any number of other places. Although raccoons remain inactive till warmer weather arrives, they do not really hibernate.

* White-tailed Deer
Odocoileus virginianus

This graceful deer is probably North America's best-known game animal. Its habit of browsing on tree twigs, buds, fruits, fresh ground vegetation, and other herbaceous foods has allowed it to thrive in many situations, even on the prairie. It is found within woodlands only where the food is not too high above its head and the ground not too shaded to support fresh growth. The edges of forests, both coniferous and deciduous, and the banks of streams are really ideal white-tail habitats.

A gland situated between the two main parts of the hoof excretes a strongly scented substance that makes the animal's footsteps easy to follow. Does are known to retrace their steps this way, as well as to track their fawns through the woods. The white-tail has three other sets of scent glands as well: one in front of the eyes; one on the outer hind limbs between the hoof and the ankle area; and one on the hind ankles. This last gland, which gives off a particularly distinctive smell, is surrounded by coarse hairs. When white-tails urinate down the inside of their hind legs, they saturate these hairs and alter the character of their scent. In this way each individual establishes its own odor.

The Columbia white-tailed deer, *O. virginianus leucurus,* is on the endangered list.

▲ Common shrew ▼ Hedgehog

Common Shrew
Sorex araneus

Shrews spend their entire lives either sleeping, mating, or feeding, the latter at least every three hours or so. Under such circumstances they cannot tolerate crowding, and have, in fact, evolved an adaptation that ensures equal space for each individual, a sort of squeaking contest. When shrews' whiskers touch, the animals squeak; and intruders usually give way to established residents. Sometimes the battle will escalate, however, and these squeaking matches, along with wriggling "exercises," tail-biting bouts, and other activities that are energetic but not physically dangerous, have earned the animals their undeserved reputation for extremely vicious fighting.

Sorex araneus possesses a long, sensitive snout ideally suited for probing about in ground litter. If insects and other foods are lacking there, the shrews surface and forage among the grass stems (they climb stems readily) and in other aboveground places. By the time they surface, of course, they are often well on their way to starving and may keel over and die at any time.

Hedgehog
Erinaceus europaeus

European hedgehogs are such effective checks on rodents and other prey animals that their presence near gardens and farms is heartily encouraged by man. When in danger, this insectivore rolls itself into a spiny ball; because of specially adapted muscles, it can remain rolled up without tiring for some time. Its natural resistance to bee and wasp toxins, as well as to the venom of snakes that it consumes, provides added protection. Winter presents no threat either, since the hedgehog hibernates in colder regions (it is the only insectivore to do so). Fat deposits usually sustain it through hibernation.

This hedgehog practices a ritual that some observers call "self-anointing": whenever the animal encounters a new object, particularly something strong-smelling, it may lick it until its own saliva is worked into a froth. Then the hedgehog spreads the foamy substance over its spines with its tongue, often doing a partial handstand at the same time. The practice is fascinating, partly because so little is understood of its purpose. One theory suggests that the saliva, having taken on the smell of the object the animal has licked, supplants the hedgehog's odor with a less personal one, and thus serves as a form of camouflage.

Old World Mole
Talpa europaea

Underground living is relatively easy for the mole when earthworms are plentiful. This insectivore usually forages for worms and other foods from within the passageways of its subterranean chambers. Those animals it does not consume right away it may paralyze with a quick bite in the head area; or it may immobilize its prey by snipping off a part of the victim's body. Once immobilized, the prey animals are added to the mole's winter store. The practice helps assure the mole of food at least once every twelve hours, a prerequisite to the animal's survival.

Old World moles use their large and rather rigid foreclaws to churn their way through forest soils. The animals are less at home aboveground, since the claws are permanently turned outward and are not really effective in helping the animal maneuver over land. The mole's marvelously luxurious fur lacks nap, and so permits both forward and backward movement through the tunnels. As one would expect, the eyes are practically nonexistent and probably only distinguish darkness from light.

Lesser Horseshoe Bat
Rhinolophus hipposideros

Before they begin hibernation in October and November, these bats accumulate the fat reserves they need by feeding on insects. Young *Rhinolophus* bats, which have relatively small fat reserves, offset cold weather by clustering in winter; the fatter adults tend to hibernate singly. *R. hipposideros* must also reduce water loss during hibernation, since it hangs with the wings wrapped around the body (and thus exposed to evaporation) instead of with the wings folded, as most other bats hang. A cave or similar hibernation site must have a relative humidity of around 95 percent, therefore, in order for *R. hipposideros* to be able to use it.

Lesser horseshoe bats are considered deep hibernators, hibernating up to three months without interruption. Even when they are not hibernating, they go into a daily torpor after the nocturnal feeding period. The horseshoe bat's common name is derived from the animal's prominent nostril flap, which appears to serve as a megaphone for sounds made during echolocation.

▲ *Long-eared bat* ▼ *Pipistrelle*

Pipistrelle
Pipistrellus pipistrellus

The tiniest and best-known of European bats is the common pipistrelle. Long ago, its erratic flight pattern earned it (and other bats) the name "flittermouse" (or *Fledermaus* in German). It begins feeding on insects at dusk, when it is still visible to the human eye, and alternates its periods of hunting with periods of rest until early morning. Then it falls fast asleep in an inverted position, often on some concealed part of a building. The final joint of the tail is prehensile, and is thought to be a significant support for the bat as it crawls up or down vertical structures.

If these bats hibernate at all, it is for a very short while and only in the colder parts of their range. Although they mate before winter, the female's eggs are not actually fertilized until the following spring, and the single young are born in midsummer.

Moisture is important to these pipistrelles, especially in hot weather when the dampness that their hibernating caves provide is not available. Often they supplement moisture from their insect diet with freestanding water, and may be seen swooping down over ponds to drink.

Long-eared Bat
Plecotus auritus

The ear flaps, or pinnae, of the *Plecotus* bats are almost as large as the animals' bodies—they are so big, in fact, that the bats fold them away when they rest. The tragus, the lobe of skin inside the pinna, is permanently erect. A *Plecotus* bat seems to use these structures together to identify the direction from which sounds bounce back to it during echolocation. What is remarkable is that *Plecotus* bats can hover among branches and pick off caterpillars and other appealing foods, leading observers to believe that these bats may be capable of detecting sounds emitted by immobile insects on the leaves of trees, and not just sounds made by moving prey. Thus they have a real feeding advantage over the faster-flying insectivorous bats. (Experiments have suggested that *Plecotus* bats emit sounds equally well from the nostrils and from the mouth. It may be that this, too, is related to their specialized hovering and feeding patterns.)

These bats belong to the diverse Vespertilionidae family, representatives of which are found in virtually every part of the nonpolar world.

Brown Hare
Lepus capensis

In the early spring as many as a dozen male brown hares may congregate in a field, where they indulge in boxing matches, chases, and other antics that probably help them establish territories. The hares seem to lose all conception of fear at this time, and people can often approach them quite easily. At other times, however, these lagomorphs are wary and quick, their only real defenses being speed and their highly effective color camouflage.

Hares generally differ from rabbits in being less prolific, and in living completely aboveground. (Certain features of the skull distinguish them further.) Whereas rabbits usually burrow, hares simply make "forms," or slight depressions in the grass, where they rest during the day. Here the does bear their young, called leverets; the newborns are fully furred and have their eyes open. Within a day they can run, and in little more than a week they are on their own. Were the offspring not so precocious, they would no doubt perish quickly in the grasp of hawks and other predators.

Lepus capensis has an enormous range over much of the Old World. It includes steppe and savanna habitats, as well as those of the deciduous woodland.

Old World Rabbit
Oryctolagus cuniculus

The rabbit's incredible reproductive rate—in some places it supposedly produces up to six litters of three to six young each between January and June—is more theory than reality. A natural control known as spontaneous reabsorption causes more than half of rabbit litters to be altered significantly. All the embryos in an unborn litter may die, in which case the embryonic material is absorbed into the doe's blood. Or the litter may be born, but there will be fewer young than usual. Female rabbits do not release eggs until after mating has taken place, and mating often occurs even outside the breeding dates mentioned above. Fertilization, therefore, is assured almost every time rabbits mate, and incredible numbers of young would be born were it not for reabsorption. Despite this phenomenon, rabbit populations often reach plague proportions where natural predators are absent.

Rabbits practice refection, the consumption of the moist fecal pellets they produce as they rest. Refection helps rabbits obtain all the vitamins they need—the pellets are composed of a vitamin-rich material produced by the interaction of digestive enzymes, intestinal bacteria, and cellulose from consumed vegetation.

House Mouse
Mus musculus

Among man's earliest rodent associates is the house mouse, which is thought to have evolved from an Asian steppe species. Its minimal water requirements, ability to tolerate extreme temperatures, and prolific breeding habits may have helped it to survive the long treks that human travelers undertook across hot lands—that is, if it was able to hitch a ride in people's belongings. Gradually it became established throughout much of the world, wherever there were sufficient grain stores and other foods to support it. Huge numbers of house mice can be a severe threat to farmers, particularly where owls and other predators are not plentiful. Indoors, house mice feed on a wide variety of leather goods, book bindings, furniture, and other foods.

Although a wild form of *Mus musculus* exists, many people have been preoccupied, historically, with the peculiarities of the more domestic form. So-called singing mice, for example, were reportedly kept caged at one time, much as canaries are today. Their "songs" are now suspected to have been a symptom of some respiratory ailment.

House mice are classified in the family Muridae, one of the largest of all mammal families, containing about four hundred species.

Norway Rat
Rattus norvegicus

Adaptability personified, Norway rats are omnivorous and omnipresent. Wherever there are cities, farms, or fields, these rodents may be found eating their way through books, trash, electrical wiring, grain, lead, soap, and concrete. Since their incisors grow some 4 to 5 inches (10–13 cm) every year (all rodents' incisors grow continually), they must be kept worn down. Gnawing accomplishes this and accounts for the billions of dollars worth of damage the rodent causes.

In addition, the Norway, or brown, rat is a notorious carrier of diseases as diverse as bubonic plague, trichinosis, typhus, and rabies (in some cases the rat's body parasites spread the disease). Fortunately, a host of predators consider this rat fine fare; otherwise its prolific breeding patterns would no doubt result in huge unchecked populations.

If there is a redeeming quality about the Norway rat, it may be that its white mutant form has been an invaluable laboratory animal. Otherwise, from the human point of view, the pugnacious Norway rat is more than a little obnoxious.

Edible Dormouse
Myoxus glis

These dormice somersault through the branches more casually than tree squirrels do. Even at night they are sure-footed. Should they misstep and fall, they invariably land on all fours in catlike fashion. They build their nests of mosses and fibrous vegetation—one for summer, a separate one for winter—in mature deciduous trees. The summer nest is usually built in a high fork or hole, while the winter nest is much lower—sometimes, in fact, in the hollows of the roots. In some locales these dormice adopt buildings, particularly barns and houses, for their hibernating sites.

The edible dormouse, the largest dormouse species, becomes extremely fat in preparation for its six- or seven-month dormancy. It is after this fat build-up has accumulated that the species is considered delectable to the human palate. No new culinary sensation, dormice of this species were popular fare as far back as Roman times.

Myoxus glis inhabits both the broadleaf forest zones of the western Palearctic and the areas around the Mediterranean Sea. It has also been introduced into Great Britain.

Northern Root Vole, or Pine Vole
Pitymys subterranus

Many people think only of moles and shrews when they think of small mammals suited to an underground life style. The pine vole, or root vole, is also adapted to subterranean activity, as evidenced by its relatively small eyes and ears, its short tail, its forefeet (enlarged to aid in digging), and its tightly packed, velvety coat, which permits movement in any direction through a tunnel.

Root voles tend to have few offspring, usually two to five per litter, a characteristic that is correlated with the female's having only two pairs of nipples. Perhaps to compensate, these voles experience a lengthy breeding season that may continue well into winter. Theoretically, as many as nine litters can be produced per year.

Thick bodies, short tails, and very small eyes generally distinguish voles from mice. Voles of the genus *Pitymys* are often classified as a subgenus of *Microtus.* Although members of the two genera differ in several anatomical features, ecological similarities betwen them are considerable. In fact, the higher degree of subterranean activity on the part of the *Pitymys* species is the only significant difference, the members of both genera being small terrestrial herbivores that burrow.

Old World Wood or Field Mouse
Apodemus sp.

Like many other rodents, wood mice are at home in a variety of habitats, including man-made structures. Where they occur on open land, they are usually outnumbered by voles, but in the deciduous forest the situation is reversed. Wherever soil is soft enough to permit it, the wood mouse burrows and nests underground. Where it lives aboveground, it follows regular passageways and corridors, but is usually much less compulsive about doing so than the voles and shrews. It is omnivorous.

The nine or so *Apodemus* species and at least one species of a similar New World genus, *Peromyscus,* share a characteristic method of self-defense known as tail-slipping. When an enemy tries to seize one of these species by the tail, the skin on the tail may come off, without causing any real damage to the rodent itself. (Little bleeding occurs, and the rodent's suffering seems minimal.) Eventually the injured part of the tail itself falls off. The loss could be dangerous to arboreal species, since they would probably rely on their tails in climbing. Wood mice, however, are primarily terrestrial, and seem to take little notice of the damage.

Old World Badger
Meles meles

Badgers excavate their homes, known as setts, in hedge-rows and the shrubby undergrowth of forests throughout most of Eurasia. These subterranean layouts may meander for as much as a mile (1.6 km). Bedding of dried mosses, leaves, and other forest vegetation is kept fresh by being replaced periodically. The badger emerges from the sett after dark, its sensitive nose testing the air carefully for unwelcome scents before it surfaces completely.

Old World badgers typically weigh up to 30 pounds (13.6 kg), nearly twice as much as New World badgers. They have large feet and a skull that is practically impervious to head-on blows. Their unhurried gait is both an expression of their temperament and an indication of their status in the animal kingdom—badgers have no enemies besides man. Yet for all their sturdiness and nonchalance, these are amazingly playful animals, and the cubs often line up for games of leapfrog. Perhaps mild climate and their efficient, generalized feeding habits (badgers have large canines, for tearing meat, and large molars, for eating vegetation) afford them ample playtime. The badger's harsh coat thickens in autumn, so that, in the more southerly parts of its range, the animal can remain active all winter. In the really cold climates farther north, badgers hibernate.

Forest Wildcat
Felis silvestris silvestris

Forest wildcats are somewhat difficult to distinguish from domestic cats, particularly since the two may interbreed. Perhaps the most reliable means of telling them apart is through the relative size of the brain case, the wildcat's being the larger. In addition, wildcats do not cover their droppings as domestic cats do. And they are all but impossible to tame.

Like the rabbit and many other species, these felids are adapted to various types of habitats. In fact, their ancestors may not have been true forest species at all, if one can judge from the fact that wildcats back down tree trunks instead of coming down head-first, as most truly arboreal species do. They also seem inordinately fond of basking in the sun, something most true forest species have not had the chance to get accustomed to. Yet where they have not been persecuted, they thrive in deciduous woodlands, where they climb with grace and speed. Like most wild felids, wildcats stalk their prey (small mammals and birds). When within a few feet of it, they rush and lunge.

The name "wildcat" is often applied to the New World bobcat, which is actually a member of the *Lynx* genus and not a wildcat at all.

Genet
Genetta genetta

Genets have a unique appearance—they look somewhat like a cross between a cat and a mongoose or a weasel. They hunt masterfully by night, approaching their prey in a near-creeping position, and in fact look almost serpentine when they flatten their bodies for stalking. Almost any small animal serves as food. After the genet pounces, it inflicts a series of bites with its needlelike carnassials. Throughout the whole episode, the genet may be purring loudly, and the hairs on its back may be erected, forming a crest.

It is not easy for the genet to find its way about in the trees at night, and various theories have been formulated as to how it navigates. There is evidence to suggest that the animals actually commit their environment to memory by repeatedly going over their entire territory one step at a time. Their white facial markings may also "glow in the dark" and so help them to identify others of their kind.

G. genetta inhabits the Mediterranean type of forest found in Spain and southern France; it also ranges over much of nonarid Africa.

Polecat
Mustela putorius

These polecats, like most other members of the mustelid tribe, survive through sheer energy, cunning, and pluck. Their long bodies and short legs provide a real advantage, since they allow the polecat to remain low to the ground and thus difficult to see as it slinks along in search of prey —rabbits, rodents, reptiles, birds and their eggs, and almost any other vertebrate it can capture. In Great Britain polecats have been known to kill as many chickens as they could, then walk away without eating a bite. In fact, the polecat was once nearly wiped out by poultry farmers. Whether or not there are forces besides greed that motivate such wholesale killing on the part of the polecat, no one seems to know; killing or storing large numbers of animals is not necessarily thought to be the polecat's usual method of predation. In one instance, however, a single polecat was found with some forty immobilized frogs in its larder. All of the frogs had been bitten in the brain. The biting habit may assure the polecat of food in lean times, in much the same way that the mole's store of paralyzed earthworms is thought to do.

This mammal produces the strongest and most pungent odor of all the European musk-bearers. Thought to mark territories, the scent is second in noxiousness only to the skunk's.

Roe Deer
Capreolus capreolus

Unlike most other European deer, the lovely roe deer are not highly social. Perhaps this is related to the fact that they inhabit dense deciduous forests and therefore do not need the protection from adversaries that more exposed species require.

These deer are not only smaller than many others, weighing only about 55 pounds (25 kg) and measuring 27 inches (70 cm) or so at the shoulder; they are also on their own annual schedule, the timing of which differs considerably from many other deer species'. Bucks grow antlers in the winter instead of in the summer, and by the middle of April they have marked their territories by rubbing their tines and heads against conspicuous trees. The peak of the rutting season falls in late July and early August, about two months before that of the red, fallow, and Sika deer. Each buck claims a single doe, or perhaps two or three, and chases her madly over a prescribed course as a part of courtship. Because of the phenomenon of delayed implantation, the twin fawns are not born until the following summer, along with the young of most other deer species.

In autumn, roe deer often experience a "false rut." This may be triggered by the light conditions, which are similar to those in spring.

▲ Giant panda ▼ Wild boar

Giant Panda
Ailuropoda melanoleuca

Giant pandas are consummate forest mammals, at home both on the ground and in trees and adapted to a diet made up almost exclusively of bamboo. Special pads on the undersides of their forefeet help pandas to grasp gently. Because of this equipment, pandas can detach bamboo leaves from their stems without bruising them at all. They may also use their front teeth to strip off the tough outer layers of a bamboo stalk, then eat only the pith. When the panda's jaw is in a certain position, the enlarged upper canine fits right into an obvious "slot" in the lower canine. This specialization, plus large molar and premolar surfaces and certain shifts of the jaw, help the panda to grind its food.

Although eating bamboo occupies at least half the panda's time in the wild, the animal is classified as a carnivore, since it does eat meat on occasion. More detailed classification varies, depending on who is doing the classifying. Some scientists consider the giant panda a bear (a member of the Ursidae family), and some consider it a close cousin of the raccoons (Procyonidae).

Wild Boar
Sus scrofa

Parties of wild boars roam the forest and other areas by night, rooting for acorns, fern roots, and a host of other, often unusual, foods. For example, wild boars seem to gravitate toward truffles, the underground fungi relished by gourmets. Rooting with the long snout aerates forest soils, but in crop lands it can spell disaster. This is one reason that the animals are so heavily hunted; another is that they provide tasty meat.

House construction is a specialty of this wild pig. The animal clips long grasses and then climbs under them. As the boar straightens up, it lifts the grasses so that they mix with the taller plants of the herb layer, forming a canopy of sorts.

The young of this species bear definite stripes extending the length of their bodies, a form of camouflage that may indicate that the animal's original habitat was a place of dappled light and shadows. Today the boar is at home in many settings, particularly those where it can wallow in the mud.

Chinese Water Deer
Hydropotes inermis

Harelike locomotion distinguishes this mammal from most other deer species. The water deer, found in eastern China, will lie perfectly still and well concealed from predators in marshy reed beds, then take off suddenly. Its ability to make a quick getaway from predators is this small deer's main form of self-defense, since its long canines are used only to combat rivals. The canines are actually mini-tusks, with a sharp edge on their inside surface, yet they do not seem to interfere with the water deer's eating. Neither sex has antlers.

This animal has impressed observers with the insulating properties of its coat. The coat permits so little body heat to escape that a water deer can lie in the snow without melting it. The summer coat is substantially lighter in weight than the winter one.

No one seems to understand exactly why the reproductive capacity of the Chinese water deer is so great— up to six offspring are born at once. Perhaps this is a hedge against the high infant-mortality rate caused by foxes, weasels, and other predators.

Musk Deer
Moschus moschiferus

Small size and a shy, retiring nature have no doubt helped this species to survive despite years of indiscriminate trapping. It is much sought after by the soap and perfume industries for the musk produced by a specialized abdominal gland in the male. Since only a small amount of musk (about an ounce [28 g]) is available from each animal, the price on the individual's head is high.

These are the most primitive of all living deer (members of the Cervidae family). Although they have no antlers, both sexes have elongated upper canines. The males' may reach a length of some 4 inches (10 cm); the females' are much smaller. The jaw of both males and females has a muscle that pulls the canines upright when the mouth is open. (At other times these canines point backward.) Although their fights rarely if ever result in death, musk deer males will slash each other with their canines in annual combats over territories. And they are almost compulsive about keeping a special marking spot within the territory scented with musk and covered with droppings.

Musk deer are found in Siberia, northern China, Tibet, and parts of northern India and the Himalayas, where they frequent both forest and brushy areas.

3

The Coniferous Forest

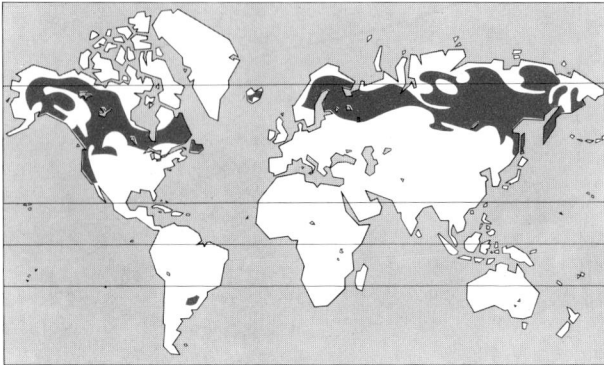

The north woods, the boreal forest, the taiga, the coniferous wilderness—all describe some aspect of the vast, primarily evergreen forest that stretches around the northern hemisphere between the tundra and 50° latitude. This is the largest forest on earth. In fact, the Eurasian taiga alone is larger than any other woodland. Partially because of its size, the taiga is still greatly undeveloped in both the Old World and the New.

This is cold country. Temperatures usually range between -22 and 68° F (-30–20° C), and the snow lies several yards deep for much of the year. The spring thaw is important to taiga vegetation, since the total precipitation in this biome is only moderate (12–29 inches [30.5–74 cm] annually). Rainfall occurs primarily during the three to five months of summer, when constant temperatures of 43° F (6° C) or warmer permit plant growth.

Characteristic taiga soil is of a type called podsol, from a Russian word for "ashes." Organic litter (twigs, leaves, dead moss, and the like) is broken down very slowly in this biome, due to the sluggish action of bacteria and fungi in cold temperatures. What decomposition does occur usually results in the formation of a thin layer of dark humus. This top layer contains acids that percolate downward with precipitation and leach out nutrient compounds from the top, then deposit them lower down. Because of this action cross sections of podsol appear dark on top, ashy gray where the leaching occurred, and red-dish-brown beneath that from iron and other relocated compounds. The iron layer in podsol sometimes develops into what is known as a hardpan. This type of soil inhibits tree-root penetration and drainage.

Acidic, nutrient-starved, poorly drained soil; low temperatures; a great deal of snow in winter; and poor to moderate rainfall in a very brief growing season—none of these conditions invites things to grow. Yet this biome is a nearly uninterrupted blanket of evergreens, which are wonderfully adapted. Spruce trees can tolerate winter temperatures as low as -40° F (-40° C), thanks to a concentrated sap, which helps protect them from desiccation. Many conifer roots are adapted to podsol through their horizontal extension just under the ground. The evergreens' conical shape provides distinct advantages in snow by allowing the snow to tumble off immediately, or to rest on limbs that grow generally downward and so lose their loads fairly quickly. Even when snow accumulates, its weight usually does not break many branches, since they are relatively flexible.

Cones, of course, are much hardier reproductive parts than the blossoms and more delicate fruits of deciduous trees. And the specialized leaves of conifers, the needles, possess waxy coverings and little surface area, both of which help to minimize evaporation. These characteristics are important in times of drought and as a protection against freezing. In harsh winter temperatures, conifers minimize needle damage by closing their stomata, the mouthlike openings in the leaf's surface through which gases necessary to photosynthesis are exchanged with the atmosphere, and through which moisture is given off in the form of water vapor. (This latter process is known as transpiration.) As soon as temperatures rise in spring, photosynthesis and transpiration are resumed full scale. This is not the case with deciduous trees, of course, which require time to produce a tree full of leaves before photosynthesis can begin.

The actual number of conifer species is small. The dominant trees in northern Eurasia are the Norway spruce, Scotch pine, and Siberian firs and larches.

Preceding pages: Black bear standing on a beaver dam at the edge of a Wyoming pine forest

(Larches, interestingly, are conifers but not evergreens—they lose their leaves in winter.) In the North American taiga, except in the humid northern coastal regions of the West, white spruce, balsam fir, and jack pine abound. Huge tracts of these and similar conifers also grow on mountains in lower latitudes, and they form interrupted bits of a woodland known as the montane forest. Like the alpine tundra, the montane forest has some distinguishing characteristics of its own. However, it is regarded here as part of the entire taiga biome, especially since the mammals that inhabit it do not constitute so specialized a group as the alpine mammals do within the tundra.

Millions of evergreens pack the land and create a recognizable green belt of a biome. Yet within the taiga, as within any major land pattern, there are zones and variations. Traveling south from the timber line for any distance, one notices the stunted and thinly spaced trees. Wind, aridity, unstable soils, and extremely low temperatures—tundra conditions, essentially—are here softened just enough to permit the beginnings of tree growth. In moving from this northern edge of the taiga into the taiga proper, one is struck by the vast reaches of similar vegetation, the characteristic firs, spruces, and other plants mentioned above. If one continues to travel south, however, the transition from taiga to deciduous forest becomes more and more pronounced. This is the southern edge of the taiga. It is hard to tell where it stops and the northern edge of the hardwood forest begins.

Even in the taiga proper the mixed stands of evergreens and deciduous trees (such as aspens) are pronounced. Much of this mix is the result of fire. Because evergreens' dry needles are particularly vulnerable to fire, lightning and other fires are rather common. Whether the results of these fires are beneficial or disastrous is still a highly controversial subject. What is incontrovertible about taiga fires is that they open up the land to sunlight and that they make nutrients such as calcium and phosphorus, which are usually locked up in the plants themselves, available to the soil. Birches, aspens, and annual weeds draw on these nutrients to get their start, and

eventually they will colonize the area. In time they will be succeeded by the evergreens, and the taiga will return.

In the interim the mixed forest provides some ideal mammal habitats. By losing their leaves annually, aspens contribute to a relatively rich layer of humus. This rich soil, in turn, unlocks nutrients that can support a wealth of shrubs. These provide excellent food for squirrels and other mammals, many of which forage in the area, then return to the conifers for nesting and cover.

Glacier-gouged lakes and peat bogs also dot the taiga. Peat bogs develop from the lakes, usually, as sphagnum moss and other vegetation growing along lake margins eventually meet other organic material and completely fill in depressions. The development of peaty soil is usually accompanied by a localized aboveground environment that contains some of the most bizarre plants on earth—the pitcher plant, for example, which traps and consumes insects. They add a flamboyant touch to the nearly solid mass of greenery that is the taiga.

Snow. It defines the northern mammals' world for much of the year, and is closely linked to the evolution of many of the more fascinating mammal adaptations. Mammals usually respond to snow in one of two ways: they may simply adjust to the white mantle, or they may thrive because of it. The snowshoe hare and the lynx fall into the latter category, primarily because of the mobility that their spectacularly enlarged feet afford them. Most other taiga mammals fall into the former group and simply "do their best" in winter, whether that means migrating, hibernating, or remaining active.

Small mammals such as shrews and mice necessarily remain active. Because of their high surface-to-volume ratio, their bodies lose heat extremely fast, and eating as often as possible is their only means of offsetting this heat loss. Since they must stoke their body furnaces almost continuously, they can hardly spare time for napping, much less for true hibernation. And migration, for a mammal this small, is out of the question.

Given their predetermined pattern of year-round ac-

Redwood forest, California

tivity, small herbivores find ground snow a blessing. Its bottom layer, known as pukak, undergoes predictable changes after it has been on the ground for some time. For instance, through a natural process known as sublimation, molecules in pukak change directly from ice (a solid state) to water vapor (a gas) without first turning to water (a liquid). This transition is brought about by heat radiating from the earth, and the process forces the vapor into the upper layers of snow. (Near the surface the vapor refreezes.) An open space right above the earth is created, a veritable indoor mall for small mammals. Because of the snow insulation above and the heat from the earth beneath, this corridor's winter temperature is both constant and higher than that of the outside air. Rarely does the pukak temperature fall much below freezing.

Although these characteristics, along with the easy access to ground food, make the pukak attractive, mall living is not without its disadvantages. For one thing, rain water may eventually freeze and form an ice crust on the snow, thereby sealing over what some naturalists think are the breathing shafts of tiny mammals. Most likely the resultant high concentrations of carbon dioxide force the animals to surface for fresh air. Once they surface, of course, they become vulnerable to hungry predators. Some of these predators, the weasels in particular, have bodies adapted for easy movement through the pukak. And foxes are credited with being able to sniff out prey that is some distance under the snow. When they locate a prey animal, they leap into the air and land on all fours. If they are lucky, the quarry will be under their paws.

Many of the medium-sized herbivores take advantage of the weighty snow that accumulates on tree branches. This snow simply bends the younger, more supple trees into an arc, and so makes the greenery near their tops available at a much lower-than-usual height. As snow piles up it also serves to raise the level of the mammals themselves, and gives them access to vegetation that, in summer and fall, is entirely out of their reach.

Large herbivores such as deer (there are more deer in this biome than in any other) are adapted to a winter diet that differs significantly from their warm-weather one. Since most of them do not paw through the snow to reach ground food, they simply eat those parts of trees (twigs and bark primarily) that they find growing aboveground. If moose can locate young trees with pliant trunks, they may rear on their hind legs and throw their entire body weight on them. Once bent or broken, the trees may yield tender foliage from their upper branches.

Winter food for many mammals, although often plentiful, is not as nutritious as a diet composed almost totally of fresh green vegetation. Consequently, those mammals that have distinctive winter diets tend to lose weight, and many are unrecognizably rawboned by spring. When spring finally comes, at least one species of deer (and probably other mammals as well) must make up their weight deficits within a certain period of time if their reproductive patterns are to be normal that season. In this respect snow actually serves as a form of birth control, since spring weight gain, essential for healthy offspring, is determined by how late in the season the snow stays on the ground. (Weight loss is not as critical to deer as it is to small mammals such as mice, which must eat constantly just to maintain body heat. The only reason they can survive winter weight loss at all is that it seems to reduce all of their energy requirements considerably.)

Another consequence of northern snow is that it keeps most mammals from moving about quickly and freely. Moose are adapted to wade through a certain amount of snow easily, primarily by virtue of their long, specialized legs. Because their legs move straight up and down, moose can travel without even dragging their hooves. Where snow is extremely deep, however, a moose may be in danger, since it can be cut and even immobilized by a hard, crusty snow cover. Caribou are tundra mammals that migrate to or from the taiga according to a combination of factors—snow conditions and availability of food. Unlike the solitary moose, caribou travel in herds, a habit that helps them maneuver in the snow by simply playing follow-the-leader. That way, only a few front runners risk cutting their legs on hardened snow.

Northern Pika
Ochotona hyperborea

Pikas are among the few animals that survive the winter by drying vegetation in the sun ahead of time. They store their food under rock ledges or, if they burrow (which the plains species do), at the entrances to their homes. Then, because foraging in the snow is unnecessary, pikas settle in for a fairly comfortable winter, though they do not hibernate or even nap for long periods at a time.

Because these lagomorphs can actually throw their voices like ventriloquists, they can be difficult to locate. They are so small that when danger threatens, they can dart into cracks and crannies among the rocks in which many of them live. Pikas differ from rabbits and hares in having limbs of uniform size, diurnal habits, and an inability to sit up on their haunches or even to hop. In addition, their reproductive rate is relatively low. The gestation period is thirty days, and two or three litters of two to six offspring each are born per year.

Two New World species inhabit the mountains of western North America; twelve Old World pika species inhabit much of Asia. The northern pika lives on talus slopes in Asia's taiga, and its range extends slightly into the tundra as well.

Wood Lemming
Myopus schisticolor

Without the moss cover on the fir forest's floor, the wood lemmings would probably perish. They eat mosses of at least one genus, as well as juniper bark and the stems of certain berries. And they build their homes in the moss that grows over the roots of birch and spruce trees.

These tiny rodents are found at fairly high altitudes (possibly up to 8,200 feet [2,500 m]) in parts of Scandinavia and northern Asia. Their populations fluctuate periodically, just as true lemming populations do. But there is some question as to whether wood lemmings migrate for any significant distance, even under pressure. Most likely they restrict their movements to their immediate locales.

True lemmings and wood lemmings both have specialized thumbs. The true lemming's thumb is actually a long, flattened claw, while the wood lemming's is more naillike, with a notch in its end. Since the wood lemming is seen very infrequently, little is known about this specialized digit, or about many of the animal's other attributes.

European Tree Squirrel, or Red Squirrel
Sciurus vulgaris

The genus *Sciurus* is composed of some fifty-five species and various subspecies. It includes both the eastern gray squirrel of North America, *S. carolinensis,* and the European red squirrel, *S. vulgaris.* Both are extremely common and familiar rodents, being predominantly diurnal and relatively fearless of people as they go about their business of searching out, eating, and burying nuts, and consuming other primarily vegetable foods.

Long ear tufts develop on this red squirrel along with its winter coat. (There are two molts a year, one in spring and the other in fall.) When the ear tufts are erected and the squirrel raises itself on all fours with its fluffy tail curled over its back, the animal looks much bigger and more imposing than it actually is. Thus its appendages serve as a form of communication, in this case a bluff.

Red squirrels' nests, usually made of twigs and mosses, are built in the forks of tall trees. Like the North American red squirrel, this tree squirrel usually "insures" itself against bad weather and danger by having more than one nest ready for habitation at any time.

Old World Flying Squirrel
Pteromys volans

Conifers provide food and shelter for these Old World rodents. At twilight they emerge from their tree-hole nests or from their mossy homes built carefully in the forks of spruce or pine. A quick leap into the air with limbs extended opens their gliding membrane (patagium). During a glide, flying squirrels may shift direction before they alight on a tree trunk, where they feed on bark and insects. Tender buds and fruits also appeal to them, even if they must climb a bit to reach them. When at rest, these rodents fold the patagium under them, and the coloration of their fur matches the tree's bark so closely that they are almost impossible to distinguish.

In members of the *Pteromys* genus, the patagium runs between the wrist and ankle only, whereas in *Petaurista* flying squirrels it extends all the way to the tail area. Most of northern Europe and Asia host the former species, while the latter is found much farther south.

Siberian Chipmunk
Eutamias sibiricus

Although they do go into a winter torpor, Siberian chipmunks do not really hibernate, as ground squirrels do. Hence the seeds that they gather from conifers are cached for winter use, not eaten to build fat reserves. They are stuffed into the chipmunks' cheek pouches, then emptied into burrows later on. Periodically throughout the winter the chipmunks wake and feed on the seeds right in their burrow chambers.

These rodents compensate for the dryness of their winter food by eating mushrooms, bulbs, birds' eggs, and other more succulent foods during warmer seasons. They sometimes damage commercial fruit trees in their search for food.

Siberian chipmunks and the chipmunks of North America's West are very similar and in fact belong to the same genus. They differ from the chipmunks of eastern North America (genus *Tamias*) only in a feature of dentition—*Eutamias* has two premolars on each side of its upper mouth, whereas *Tamias* has just one.

* **Brown Bear**
Ursus arctos

Brown bears, grizzlies, Kodiak bears—all are classified in the same genus and species by many scientists, though there may be a difference of several hundred pounds among the various forms. What they all have in common is, among other things, an extremely generalized diet, especially for a large carnivore. Although some individuals may eat meat primarily, most consume an amalgamation of honey, fruit, berries, bulbs, fish, carrion, and even an occasional deer.

Brown bears mate in late spring, and the females bear one to four cubs while still in their winter semidormant state (usually in January or February). Though the mother may weigh close to 1,750 pounds (795 kg), the cubs weigh only 1 to 1½ pounds (454–680 g) each, and measure no more than 9 inches (23 cm) in length. The very fact that they are born when the mother's heartbeat and breathing are considerably slowed is remarkable, especially when one considers that the tiny offspring grow to a weight of some 50 to 60 pounds (23–27 kg) within their first year of life.

One subspecies of this bear is endangered, *U. arctos nelsoni,* the Mexican grizzly. In fact, it may already be extinct, apparently having been poisoned and hunted out of existence in the last three or four decades.

Red Fox
Vulpes vulpes

The extremely adaptable red fox ranges over an enormous area—most of Eurasia and North America as well as Australia, to which it was introduced in the late nineteenth century. This canid seems to thrive in almost any environment, but prefers a mixed habitat with some cleared land. Human settlements are attractive to it, and fox hunting is still a popular sport in many places. Stories of the fox's wily ways are perhaps even more popular: this species is the Reynard of folklore and the fox of Aesop's tales.

A varied diet that includes rabbits, small rodents, birds, insects, and fruits allows considerable leeway in this fox's life style. No doubt it also contributes to the species' success, as well as making the red fox a significant regulator of prey populations.

Red foxes are not always red—they exist in red, black, and cross-color phases. Some members of this genus can be distinguished from other foxes by the pupils of their eyes, which become elliptical in bright light; and noticeably active scent glands in the feet, tail, and anal region characterize some of the species. The glands produce a strong, pungent smell, especially when the fox is under pursuit or otherwise frightened.

Mating usually occurs in January and February, and competition for mates can be fierce. A repertory of yaps, grunts, and snarls helps the males and females to find each other and warns unwanted individuals to stay away. Gestation is fifty-two days, and the vixen and her three to six pups usually confine themselves to a den area until well after weaning. Both parents participate in feeding the older pups small mammals, though the male often regurgitates his food to the vixen outside the den. By autumn the family unit disbands.

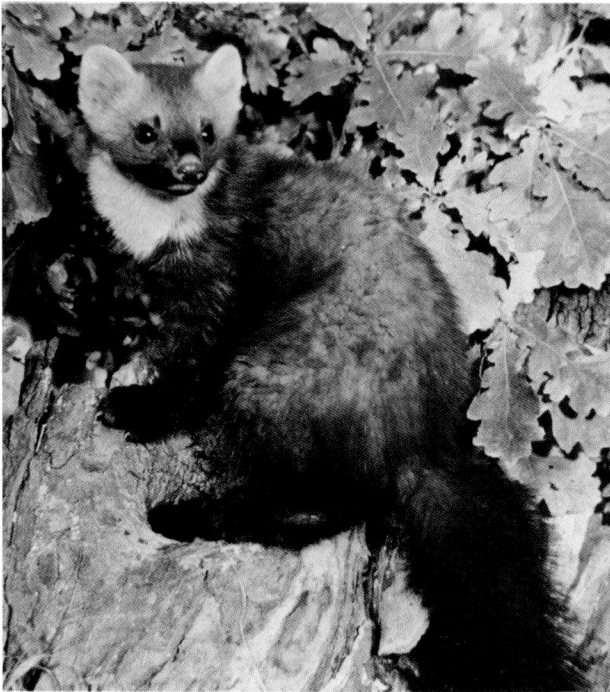

Pine Marten
Martes martes

Pine martens are specialized mustelids—they live in trees and prey on squirrels; yet they also take other animals, including mice, fishes, and even deer. By night they chase squirrels over branches at amazingly high speeds, or they may sneak up on them in their nests. They bite their small prey in the neck; in bringing down large animals they may go for the carotid artery. The pine marten is such a skillful hunter that gray squirrels did not flourish in the British Isles until the pine marten and the polecat became scarce there. (The pine marten's range includes most of Europe and part of eastern Siberia.) Supplementary foods include seeds, berries, and birds' eggs. The martens stash the latter in the mossy coniferous-forest floor, preserving them for as long as several months. They also consume slugs, after rubbing off their slime.

These mammals have close cousins in the New World—the American marten, *M. americana,* and the fisher, *M. pennanti.* They, too, fill the niche of arboreal carnivore, as does another close cousin, the Old World sable, *M. zibellina.* The sable's range has been greatly diminished by earlier trapping for the opulent fur. Like minks, sables are now raised commercially in special quarters, from which some are being reintroduced to the wild.

European Mink
Mustela lutreola

The mink stops at nothing to trap its prey. It goes right into water voles' burrows to make a kill, and may even take over its victim's home for its own use. European minks frequent stream banks throughout parts of Russia and Europe, where crayfish, frogs, and fishes are also available. And although they are not as quick at it as weasels, they kill poultry when they can.

Snow does not seem to bother the mink a great deal, and in fact the young, called kits, slide playfully down snow banks. No doubt the mink's famous coat keeps the animal warm. The European mink's coat is actually less luxurious than the American mink's, and in some parts of the Soviet Union the American species, *M. vison,* has been introduced as the European species has declined in numbers. Today, however, most minks used commercially are raised on special fur "farms."

Mating occurs in February and March. Because of delayed implantation of the embryo in the uterus, which is typical of mustelids, the two to ten kits may not be born until nearly three months later.

* Lynx
Lynx lynx

Enormous feet, ear tufts, and a bobbed-off tail characterize this cat. It is the only feline that is restricted to the northern woods. Slightly differing species of this genus live in both the Old World and the New. The latter, called the Canada lynx, has been the subject of many studies, now classic, involving its chief prey, the varying hare. These studies have shown lynx populations to fluctuate in nine- to ten-year cycles that are parallel to the hare's. In the Old World, too, hares form the staple food of this carnivore, yet no one completely understands these animals' cycles. The caracal, also known as the desert lynx, is sometimes classified in its own genus, but is very similar in appearance to the lynx.

The lynx is obviously well outfitted for its cold range. Its large "snowshoe" paws give it maneuverability, and its magnificent coat provides insulation. The pelage, however, has proven a liability, since commercial hunting has helped to reduce the Old World lynx's original range until today it encompasses little more than the Russian taiga. Development of forest areas has also had its impact: the lynx has been forced to adapt to cultivated land, which it can do so long as there is sufficient prey nearby. The Spanish lynx is endangered, due to its being persecuted for supposedly killing livestock.

Varying hare in summer (l) and winter coats ▲

Pygmy Shrew
Microsorex hoyi

The fretful shrew hasn't time to sleep away the winter. It must always be about hunting (nearly any kind of food will do) and breeding (two or three litters of three to ten each are born per season). The high-strung shrew cannot keep up this frantic pace for very long—it dies after only a year or two.

Pygmy shrews are the smallest American mammals, weighing between .08 and .12 ounces (2.3–3.3 g), or about as much as a dime. It is remarkable that they can endure northern forest conditions; undoubtedly they would be unable to survive without tunneling underground or under the snow, where the microclimate is fairly agreeable. The holes they make in leaf litter during their travels are narrower than a person's little finger.

Watching the tiny shrew dart about or battle for food requires close attention, so quick are the animal's movements. Squeaks and twitters accompany a rush here, a jerk there, and the shrew may disappear from sight.

Varying Hare, or Snowshoe Hare
Lepus americanus

This mammal's coat, which lightens and darkens seasonally, provides an ideal camouflage. Apparently the color change is triggered by the number of hours of daylight. In autumn and winter relatively little light strikes the animal's eyes, and this reduced light somehow cuts down on the production of brown pigment. In spring and summer the process is reversed.

Long, stiff hairs grow on the sides of this mammal's hind feet in autumn. These are the "snowshoes" from which the hare gets its name. They provide good insulation from the cold as well as extra surface area, which keeps the animal from sinking in the snow.

Extreme population fluctuations, still not fully understood, are characteristic of this lagomorph. About every ten years incredible numbers of varying hares are in evidence: up to five thousand hares per square mile (2.6 sq km) have been recorded. The numbers decline gradually for about two years following this peak, and then, suddenly, the animals all but disappear. (There may be, in fact, no more than about thirty individuals per square mile at the lowest point in the cycle.) This condition persists for about three more years, then gradually the numbers of varying hares begin to increase. Populations of predators such as lynx follow a generally parallel pattern.

Aplodontia, Mountain Beaver, or Sewellel
Aplodontia rufa

The unusual musculature of its head makes this mammal seem like a walking fossil. It is the most primitive of living rodents and the only representative of the North American mammal family known as Aplodontiidae. The aplodontia shares certain characteristics with the lagomorphs. It ingests its moist fecal pellets, as all the lagomorphs do, and it harvests grasses and some other tender vegetation for drying in the sun, as pikas do.

Known also as sewellels, aplodontias thrive in damp surroundings. In fact, their range is limited to the moist evergreen forests along North America's western shore. Excavating wet soil is a favorite activity of this burrowing species, and it usually chooses to travel through damp routes rather than dry ones. Even in the spring, when the lives of other mammals are disrupted by the washing away of their homes, the aplodontia thrives.

North American Red Squirrel
Tamiasciurus hudsonicus

These red squirrels take no chances with the weather—they have three or four nests for insurance. The winter ones are usually holes in tree trunks or, alternately, soundly built structures of leaves and twigs located in the most protected parts of trees. Burrows are used also, as well as loosely constructed summer nests. These squirrels' tendencies with food are similar: hoarding is the order of things, and caches of up to 150 pine cones per individual are not at all unusual. The cones are usually clipped and stored in a damp spot while still quite green. Thus they remain closed until the squirrels can get around to opening them and extracting the seeds. Should snow cover the caches early in the season, the red squirrels tunnel right through to reach their store.

The western red squirrel, *T. douglasii,* is also known as the chickaree; the northern and eastern red squirrel, *T. hudsonicus,* is the more abundant of the two species. Both are chatty denizens of North America's evergreen forests that take no trouble to conceal their whereabouts.

Beaver
Castor canadensis

The beaver is adapted to change its habitat. Beaver dams dot waters across North America, and are scattered over parts of the Old World as well. The dams are made of logs, stones, and roots plastered with mud. The ponds that result serve as storage places for aspen and willow branches and other winter foods that the beavers anchor in the bottom mud. They construct a lodge in the midst of the pond or sometimes on a bank. This is a colony's home for several years, and the members swim in and out through their underwater entrance year-round, even when pond ice is thick.

Perhaps no other well-known animal has so many characteristics that capture the human imagination. The slapping of the beaver's paddle-shaped tail on the water is a widely heard alarm; its incisors, which grow throughout its life, are perfect examples of a rodent's chisellike tools; its delicate forefeet are adapted to handle wood and other objects gracefully and deftly; its well-oiled fur provides prime pelts, and for years fur trappers contributed heavily to reductions in the beaver's numbers; and its very size is enough to startle people who see the animal for the first time. At about 44 pounds (20 kg), it is the second largest rodent in the world, surpassed only by the capybara.

Deer Mouse and White-footed Mouse
Peromyscus sp.

Deer mice and white-footed mice are all but impossible for the layman to tell apart. They are among the most abundant of mammals, occurring in the New World in almost every dry habitat in the northern hemisphere and even into South America. This is curious, since their long ears and tail would seem to lose heat rapidly and therefore not be well adapted to colder climates. In the deciduous forests of eastern Canada white-footed mice compensate for their long appendages by huddling with others of their kind in winter, a practice that conserves their body heat. The huddles have a social code all their own: the dominant members of the colony take their place in the center, where the warmth is greatest. Members of these winter groupings forage together for food, although both white-footed mice and deer mice store seeds in their quarters long before winter is upon them. Deer mice even separate the seeds according to type.

These rodents build their nests aboveground, in any number of sites from tin cans to tree cavities. Even boots and mattresses have housed members of this rodent genus.

▲ *Porcupine* ▼ *White-footed mouse*

Porcupine
Erethizon dorsatum

New World porcupines, noted for their effective self-defense, are also excellent travelers. Climbing trees is second nature to them, which it is not to their Old World counterparts such as the brush-tailed porcupines. And swimming seems effortless, thanks to the buoyancy that their hollow quills provide. Their mobility gives porcupines access to a wide variety of foods, from tree bark and foliage to water-lily leaves. And it is no doubt one of the main reasons they can stay active year-round in deciduous and coniferous forests.

Porcupine mating is accompanied by much ceremonious humming, dancing, and nose-rubbing. A clap on the back can send a potential partner reeling, yet after prolonged foreplay, coupling normally proceeds without injury. Female porcupines may have several mates each season, and bear one offspring at a time.

Black Bear
Euarctos americanus

Grasses, clover, bark, berries, roots, bulbs, small mammals, sheep, even other black bears—all these and more have a place on the black bear's menu. The food must go a long way, since black bears eat enough during the warm-weather months to see them through the winter. They do not really hibernate, but they do go into a deep slumber. During this time their alimentary tract ceases to function, and a fecal ''plug'' is formed. When the bears emerge in spring, they expel the plugs and soon begin eating again. Tourists in the national parks of the American West find panhandling bears rather commonplace, although in the wild these mammals will climb a tree rather than confront a human being or other potential enemy, particularly the huge grizzly.

Black bears begin to mate soon after they emerge from their dens in spring. For five to six months thereafter, however, the fertilized eggs remain inactive. They begin to develop only during the ten weeks or so prior to the tiny cubs' birth in January or February. This phenomenon, known as delayed implantation, occurs in several mammals and is thought to be a form of population control.

* Cougar, Puma, or Mountain Lion
Felis concolor

Cougars (also known as pumas and mountain lions) at one time ranged over more of North America than any other mammal. Now the eastern cougar, *F. concolor cougar*, is endangered, as is the Florida cougar, *F. concolor coryi*. The few cougars that survive in North America live primarily in the wilderness areas of the West, where they are often forced to prey on domestic animals instead of deer, their preferred staple. The narrowing of their range was brought about by a severe decline in the numbers of deer; and the deer population declined because of the wholesale clearing of forest land, prime deer habitat.

Although quite ferocious in captivity, the cougar is rather furtive and prefers to be left to itself in the wild. During breeding season the female performs a double duty: she not only feeds her kittens, but also must defend them from their father—he will more than likely eat his kittens, given the chance. This habit is actually rather common among animals, and noticeably so among cats in America. Perhaps it is a form of population control, since cougars are known somehow to keep their numbers at remarkably stable levels.

Elk, or Wapiti
Cervus canadensis

Elk stags bear a spectacular headdress of antlers, which serve them well during the autumn mating season. Bugling loudly, rival males battle one another for the cows, shoving with their antlers, but rarely locking antlers as other deer do. The prize for the winner of a spar is an entire harem, up to fifty or sixty females. Some naturalists believe that the females actually choose their master by simply gathering near him. Others think that the males coerce each female in turn.

These mammals live in a variety of habitats, from alpine meadows to coniferous forests and even open prairies. Regardless of where they summer over, they migrate into the canyons and other warmer areas as winter closes in. Small bands may join together during the journey until sizable herds are formed. When the browse runs low in the valleys and the elk are starving, they may invade orchards or, in desperation, seek out hay. Spring brings a reversal of movement, and females bear their calves in meadows well along the route up the mountains. Calves often survive the threat of wolves and bears simply through their mothers' determination, since the males leave their families to fend for themselves.

This deer, the largest in the world except for the moose, is very similar to the red deer, which is the prototypical deer of the Old World. A longstanding confusion over its common and Latin names still persists, and *Cervus canadensis,* also known as *C. elaphus,* is often referred to as both the elk and the wapiti.

Moose
Alces alces

Unique in appearance and in almost every other respect, moose are the largest deer in the world, standing up to 6 feet (1.8 m) at the shoulder. The shovellike antlers do not develop until the animal's fourth or fifth year. Fully developed, they may measure almost 6½ feet (2 m) across. The antlers can be a liability: during autumn rut, competing bulls have been known to immobilize themselves by interlocking antlers. Completely helpless this way, the bulls become easy prey for wolves and bears.

These leggy waders have easy access to water plants growing in summer ponds. They dive to reach submerged food, and can swim up to 35 miles per hour (56 kph) if they have to. Even kneeling on land to reach low-growing summer vegetation presents no problems, and unless crusty snows cut their stiltlike legs, moose can get about easily in winter, too. At that time moose grow a blue-tinged undercoat of kinky hairs, a good insulation against blizzards.

In the Old World this mammal is known as an elk, in the New World as a moose. It is found outside some cities of the far North, where it pursues its solitary life rather quietly. Once alerted by the breaking of twigs or the smell of people, it will flee.

4

The Tropical Grasslands

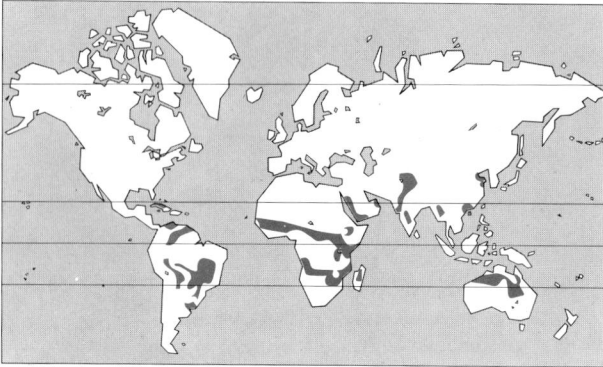

"Savanna" is the name given to the large areas of grassland found between the Tropic of Capricorn and the Tropic of Cancer. The term originally implied an area with only a few scattered trees. Now, however, it includes grasslands that are moderately or even heavily treed.

The features that distinguish this type of grassland from that of the temperate zone are the relatively constant and high year-round temperature (around 70° F [21° C]) and the extremes of wetness and drought that characterize summer and winter respectively. A summer rainfall of 58.5 inches (150 cm) is not uncommon, and is typically followed by drought severe enough to limit tree growth. The drought, in turn, if prolonged, is often followed by fire. Evaporation is usually rapid. Since the vegetation must be able to withstand such extremes, it is no wonder that only a relatively few species of trees and grasses can survive.

By far the largest savanna in the world is found in Africa, where it is wrapped almost completely around the tropical forest that lies at the center of the continent. Savanna occupies more than one-third of Africa's land mass, in fact, and it also occurs in the western portion of Madagascar.

The grasses often referred to as elephant grass may reach a height of 10 feet (3 m). They tend to dominate the African savanna, along with the baobab tree and the flat-topped acacia trees. Both of these trees have evolved

into fire-resistant species, and the trunk of the baobab into a giant water-storage vessel. The acacias, in particular, should be appreciated for their integral role in this ecosystem. Because their fruits fall to the ground during the dry season, few if any of the pods are distributed by water. Instead, their rich pulp is eaten by various types of antelopes, which pass the seeds through their digestive tracts and out in their dung. Usually the seeds stand a good chance of germinating wherever the dung is deposited (often in the grass-free clearings where animals congregate). Thus the acacia seed's distribution problem is solved through an adaptation that provides food for hungry mammals during drought.

Like Africa's savanna, South America's has its characteristic features. In the wetter savannas, called llanos, floods often inundate the low-lying land. Although they may limit or altogether preclude tree growth, as the floods retreat they leave a legacy of rich vegetation that is often eaten by domestic cattle. The drier savannas, or campos, occur at higher elevations, and are usually punctuated by scattered trees.

Northern Australia has some savanna, which varies considerably from one region to another. In India the biome is a man-made one, existing where woodland once did before it was cut down. India's situation introduces the whole question of the origin of all tropical grasslands, a question that has no clear-cut answers even now. Many scientists believe that the savanna is climatically determined, just as any other biome is. Yet there is no doubt that in some places where tropical forest has been cleared or burned, it has been replaced by savanna. This is true of well over 1,750,000 square miles (4,550,000 sq km) in Africa, as well as of portions of India.

The one fact that is unmistakably clear is that regardless of how it originated, the savanna and many of its species are fast disappearing, primarily because of human pressures. The situation is probably most devastating in Africa, where poaching is widespread. Human populations continue to encroach on what is historically wild animals' land, and domestic cattle and sheep have been al-

Preceding pages: Lion moving across savanna, Kenya

Long-eared Elephant Shrew
Elephantulus sp.

Elephant shrews and aardvarks share a fondness for termites. This tiny insectivore's elephantlike snout is unusually sensitive to smells and can easily discern a termite's whereabouts. The shrew unfolds and extends its tongue to make a capture. The process is extremely quick, and the elephant shrew repeats it many times during the course of a typical day, storing termites in its cheek pouches and eating them later on.

These are such tiny mammals (they weigh only around 1–1¾ ounces [30–50 g]) that some of them actually shelter in cracks in the soil. Or, since they are not particularly skilled at burrowing, they may rely on the abandoned homes of other mammals for shelter.

Until recently, it was thought that elephant shrews hopped on their hind limbs only. However, it has been discovered that the animals move on all fours at all times, even when they leap, which they do with great agility. The drumming of the hind feet produced by some species may be instrumental in locating prey, though no one seems to understand exactly how.

Vervet Monkey
Cercopithecus aethiops

Like baboons, vervet monkeys have evolved to a stage at which they no longer need deep forests in order to survive. Like baboons, too, vervets are extremely social and follow rigid hierarchical lines. The two adaptations do not occur together by chance, since animals as small and vulnerable as these could hardly make it alone on the open savanna. Regardless of possible insurrections, therefore, vervets do all they can to maintain the social status quo.

Of all their social activities, grooming is the only one that cuts across the lines of rank. Any vervet can groom any other vervet, and obviously the monkeys enjoy it, if one can judge by the expressions on their faces. Infants can get away with nearly any kind of social "insult" until they are about eighteen months old—it is almost as if they are granted some sort of special dispensation until that age. This permissiveness also binds troupe members together by focusing attention on the young and relatively helpless. In addition, it aids in assuring the continuation of the species.

limited to herbivores and baboons, but is evident also in the prides of lions and the hunting packs of hyaenas. These two animals represent the other side of the African wildlife picture, since, along with a few other species, they prey and scavenge on the multitudes of herbivores. There are relatively few carnivores in any ecosystem. Yet their means of living is essential not only to them, but ultimately to the survival and propagation of healthy herbivore populations: the predators often prey on the weak and sickly individuals first, and also serve as regulators of population growth. Thus, their role is actually one of maintaining balance in the community.

Being at the top of the food chain, lions socialize for purposes other than defense. Their "extended family" seems to be the most efficient unit not only for cooperative hunting, but also for the rearing of young. Cubs are taught numerous games that help them develop their forepaws and legs in preparation for future kills.

Cheetahs, even more than lions, rely on their powerful leg muscles when they hunt—they have nonretractable claws that help them grip the earth, and can sprint up to 70 miles per hour (113 kph). But their stamina is limited, and they cannot sustain their speed for more than about 500 yards (457.5 m). Thus, if they cannot bring an animal down early in the chase, they lose it altogether.

The cheetah, the lion, and the leopard can all consume enough meat from one kill to sustain them through many hours, even several days, of resting. What a contrast this is to the nearly constant grazing and browsing of the herbivores!

The final mammalian link in the savanna food chain is the scavengers—hyaenas, jackals, and African wild dogs. They serve as the savanna's clean-up crew. Without such carrion-eaters, the savanna would be covered with bones and decomposing carcasses; with them, it stays at its peak for the entire wildlife community.

Chacma baboon ▲

Baboon
Papio sp.

Baboons are savanna-dwelling primates that have evolved one of the most highly organized societies of all mammals. The key members of their social groups are the dominant males, which defend the troupes against enemies, conduct feeding expeditions, and fertilize a high percentage of the receptive females. Except in the case of hamadryas baboons, there is more than one dominant male in each social group, which usually has some twenty-odd members, sometimes more.

Females, juveniles, and subordinate males reinforce the hierarchy by constantly directing ritualized gestures toward the older individuals. They may approach in submissive postures, for example, or they may grimace at their superiors. The social order can be observed particularly well when a baboon troupe is on the move. The young males stay in front, followed by the females and offspring. In the middle of the troupe are the dominant males, with more females and young. More subordinate males bring up the rear. This arrangement assures an ever-ready supply of males on the outside, should their defense be called for.

Baboons need trees to survive, but they no longer need entire forests. They sleep in trees by night; by day they drink and search for their varied foods.

African Mole-Rat
Tachyoryctes splendens

African mole-rats strongly resemble New World pocket gophers of the family Geomyidae; the main difference between them is the absence of cheek pouches in the mole-rat. Both animals mound up soil near the entrances to their burrows, and both burrow exceedingly well. (They occupy the same ecological niche.) Mole-rats' limbs are short but well developed for digging, despite their lack of long claws, and their bright orange teeth cut right through roots and other underground obstacles.

Hygiene is not one of this rodent's major priorities. Latrine areas, food-storage areas, and sleeping areas are all quite close to one another in the mole-rats' nesting chambers. However, this apparent sloppiness may serve an important purpose, since the heat produced by decomposing wastes keeps the nests moist and warm.

The often-photographed naked mole-rat, perhaps the least handsome of all rodents, is classified in a separate genus, *Heterocephalus,* and not with *Tachyoryctes.*

Cane Rat
Thryonomys sp.

Cane rats are among the largest of all African rodents, sometimes reaching a length of about 23.5 inches (60 cm). They can be heard chewing noisily on the grasses in sugar-cane fields. In fact, they can be a menace in sugar plantations unless pythons, mongooses, leopards, or other predators help to keep them in check. Pythons are often protected to do this. Although the cane rats' pelage resembles that of the porcupines, the spines are not really thought to offer much protection. Heavy grass cover and reeds usually provide ample space in which to hide, but occasionally a cane rat is forced to dash from its shelter. It usually heads for water, where it feels quite at home.

Many native African peoples find the meat of the cane rat desirable, but are not inclined to eat rats. Since the cane rat's most "ratlike" characteristic is its tail, rat hunters ingeniously amputate the tails as soon as possible after capture. Once this is done, the hunters and others seem to enjoy eating the cane rat as if it were any other food.

Zebra Mouse, or Striped Grass Mouse
Lemniscomys sp.

When frightened, zebra mice (also called striped mice) hop straight into the air before they scurry away. They normally inhabit grassy nests of their own construction; when fire sweeps through their area, however, they are forced to seek shelter in other mammals' burrows.

No one seems to understand how these rodents can maintain a diurnal life style, as they do, without perishing in extreme heat and sun. It is known, however, that they have a black membrane covering their brain case, which may limit the amount of radiation the animals receive.

Various types of habitat serve as home for the six species of zebra mice, from swampy grasslands to true bush country. They feed on a wide variety of vegetable foods. Though as many as a dozen offspring per litter are sometimes born, the more usual number is between two and five.

Ratel, or Honey Badger
Mellivora capensis

Not all savanna carnivores are large, as evidenced by various small members of the dog and cat families such as bat-eared foxes and servals. And of course there are the mustelids, which are represented most conspicuously, perhaps, by the zorilla and the honey badger. Honey badgers, which get their name from their liking for honey, have developed an unusual association with a bird known as the honey guide. If this bird locates a bees' nest, it will alert nearby mammals with its distinctive call. Those mammals that follow the bird's call (the ratel is among them) are led directly to the nest and share the feast.

The ratel must do its part of the work, however, and this consists of digging out the nest. In fact, there is little for which a ratel will not dig, given the chance, and even less that it will not attack. During the breeding season, particularly, it frequently charges and bites some of the larger ungulates so severely that it may cause them to bleed to death. No doubt the ungulates somehow threaten the ratel. Its own skin is tough enough so that it is almost impervious to injury, especially since the ratel can twist its body around inside its loose skin as if the skin were barely attached. In this way the ratel can defend itself by biting an animal that already has the mustelid in its mouth.

▲ African wild dog ▼ Ratel

African Wild Dog, or Hunting Dog
Lycaon pictus

Hunting dogs are among the most maligned of wildlife because, like wolves and dholes, they hunt in packs and employ what, by human standards, seem to be cruel methods of killing. Wild dogs can see potential prey while quite some distance from it. They do not assume a hunting posture, however, until they are within about 400 or 500 yards (366–457.5 m). Then they lower their heads, flatten their ears, and wait for the prey animals to run. The running seems to trigger the dogs' pursuit, and they follow the leader of their pack toward its chosen victim until, together, they bite it till it loses strength. Then they pull it to the ground or simply wait for it to fall.

Wild dogs have evolved a way to ensure food for the entire pack, including the young. In response to ritualized begging, the dogs that eat first regurgitate food to the animals that did not participate in the kill, and this second round of consumers, in turn, may regurgitate it to the pups. In this way the same food may pass through several different dogs before it is discarded.

Although they begin to eat regurgitated meat when only two weeks old, the pups (there are two to twelve in each litter) nevertheless nurse for about three months and will suckle indiscriminately from any lactating female.

Spotted Hyaena
Crocuta crocuta

Hyaenas occupy a very special position among savanna wildlife. Their jaws, the strongest of all living carnivores', are equipped with rather blunt teeth that allow the animals to eat fresh meat and even the large bones of carrion (hyaenas both hunt and scavenge). Their hindquarters, however, are so weak that the animals cannot run very well, but seem merely to shuffle along in comparison with cheetahs and other swift species. Therefore, they compete with those animals by organizing into hunting packs and often by culling the slower individuals in a herd. These may be the very young or old, or the wounded or sick individuals. Hyaenas also follow vultures to food, and will take a lion's wildebeest or zebra if they can. They may produce their famous "laugh" after each kill or takeover, and through their vocalization inadvertently attract other predators. The hyaena's laugh, actually an indication of the animal's excitement, is heard frequently during the breeding season.

* Leopard
Leo pardus

Leopards are the essence of solitary stealth. They often hunt from trees, either pouncing down on unsuspecting antelopes or occasionally taking monkeys right in the branches. Their strength is so great that they can drag an entire antelope carcass up a tree, where the kill will be secure from other hungry predators such as hyaenas. There, after wedging it into a firm position between the limbs, they feast. As many as forty different species make up the leopard's diet, and the animal is so intelligent and adaptable that it can live in all kinds of habitats, from rain forests to savanna.

Today, five subspecies of this splendid cat are endangered, and others may well follow suit. Their plight is the result of increasing human encroachment into leopard territory, which forces certain leopards to hunt domestic species in order to survive. When they do, of course, they incur the wrath of ranchers, to whom they are very vulnerable. Even more tragic is the fashion trade's demand for the leopard's fantastic coat. This demand has kept many a poacher employed, a state of affairs that seems particularly unjust considering the nobility of the animal so thoughtlessly killed.

* Lion
Leo leo

Lions are among the most gentle and affectionate of all animals with their offspring. A cub's training may last up to eighteen months, and is usually supervised by assorted "aunts" and possibly other members of the pride (the lion's extended family unit), as well as by both parents.

Adult lions engage in distinctive and elaborate signaling, from marvelously varied facial expressions to head rubbing and nuzzling. They communicate over long distances via their well-known roar, the sound of which carries up to 5 miles (8 km).

Lions are perhaps best known for their effective communal hunting. They do what stalking they can in the open spaces of the savanna, two or more of them often setting up an ambush for an unwary wildebeest, zebra, gazelle, or other prey. A female usually makes the actual kill. Lions' carnassial teeth are the ultimate carnivore weapon. They act in conjunction with each other in much the same way that the blades of a pair of shears do, and they can cause almost instant strangulation when thrust into the neck of a victim. A lion weighs so much (up to 495 pounds [225 kg]) that it can crush the spine of its prey simply by leaping onto the animal's back.

The Asiatic lion, *Panthera leo persica,* is endangered. (*Panthera* is another name for the genus *Leo.*)

* Cheetah
Acinonyx jubatus

Cheetahs, the sleekest and most long-legged wild cats, are the fastest mammals on earth. They hunt by stalking their prey or by chasing it in the open. In the latter case a cheetah usually singles out its victim, often a Thompson's gazelle, and chases it at full speed (up to 70 mph [113 kph]) for about 500 yards (457.5 m). After running this distance, it is exhausted. Therefore, if the gazelle manages to dodge suddenly, or can outlast the cheetah, it will probably escape. If not, the cat knocks it down from the rear, then strangles it.

Since cheetahs do not live in large, well-organized societies, cubs are left alone while the mothers hunt. And since they do not climb trees, they have no particular spot to take a carcass. So they must eat quickly, before hyaenas, vultures, or lions claim their kill. Or they may drag the carcass into some sort of meager cover and share it with the cubs. Perhaps because they are rather nervous at being so vulnerable after a kill, they are messy eaters, consuming only the solid meat of their victims. They will, however, drink the blood of their prey, especially when freestanding water is scarce.

The Asiatic cheetah, *A. jubatus venaticus,* is endangered—its habitat has been severely impinged on, and the animal itself has been overhunted.

African Elephant
Loxodonta africana

African elephants are almost always on the move—this is not surprising, since they must find and consume enough vegetation to support their weight of up to 8.25 tons (7.5 m t). They help convert woodland to savanna by eating a wide variety of plants, including trees which they may strip of bark or which they may uproot altogether in order to obtain only a few leafy twigs. Elephant food is so rough, in fact, that the animal's molars are worn out and replaced five different times during a normal life span. Although their foraging is wasteful, elephants do provide a trail of food for smaller animals. And they are the only mammals adept at digging for water during times of drought. Their own prodigious moisture requirements necessitate their digging, but the digging in turn benefits much savanna wildlife, since deserted holes are used by rhinos, gazelles, and many other mammals.

Their nomadic life style is also a social one, elephants being among the most sensitive and communicative of animals. Large flaplike ears detect the faintest sounds produced by a fellow's mouth, its trunk, or even its stomach. The ears' position can also indicate social acceptance or rejection, and they help regulate the animal's body temperature—as blood from the elephant's body passes through vessels in the ears, it is cooled several degrees. The trunk is even more versatile—it is used to forage, to caress during courtship, to discipline the young, and even to determine conditions such as air movement and the direction of certain scents.

Poaching for the enormous ivory tusks has caused elephant populations to decline drastically. The future of the magnificent animals is bleak, in fact, because of man's encroachment on the large foraging area required for their survival.

Aardvark
Orycteropus afer

Aardvarks are one of the few mammals to feed almost exclusively on ants and termites, hordes of which are found throughout much of arid Africa. After the aardvark locates its food (presumably by smell or hearing), it uses its strong, clawed forelimbs to rip open the nest. Then it extends its long, sticky tongue inside to reach the insects. (The teeth of adult aardvarks are located toward the rear of the mouth and do not interfere with the action of the tongue.) Its thick skin protects the aardvark from the bites of swarming termites.

Burrowing comes easily for the aardvark, which may excavate simple homes or more elaborate ones that extend for 400 or 500 square yards (334–418 sq m) and have up to thirty entrances. Night brings this slow-moving "earth pig" aboveground. ("Earth pig" is the literal translation of *aardvark*, which is an Afrikaans word.) After it has abandoned a burrow, snakes, bats, ground squirrels, wart hogs, or other creatures may move in. Thus the aardvark contributes significantly to savanna housing and, therefore, to the biome's total balance.

Rock Hyrax
Heterohyrax sp.

Before the Miocene epoch, these mammals were plentiful and widespread over much of the savanna. As ungulates gained predominance, however, they drove the hyraxes into rocky regions and forests. Although hyraxes have managed well enough in these habitats, they are handicapped by their inefficient method of feeding. In order for its molars to do the actual work of cropping vegetation, the hyrax must turn its head to the side and open its mouth extremely wide. This style of eating, plus its short legs, its unstable body temperature, and its total inability to burrow, has certainly stood in the way of the hyrax's regaining its original position of dominance.

Although taxonomists have assigned them their own order (Hyracoidea), hyraxes are more closely related to elephants than to any other mammals. The similarity is obvious in the hyrax's flat toes and its upper incisors, which are rather like rudimentary tusks.

▲ Grant's zebras ▼ Black rhinoceroses

Grant's Zebra
Equus burchelli

Rarely does one see a single zebra. Representatives of the three species often intermingle as they wander in search of good pasture and approach watering holes, places where prey species are particularly vulnerable. They frequently associate with wildebeests, giraffes, and many other species. Such clustering no doubt works to each animal's benefit, for as soon as one species, through its superior sense of sight, sound, or smell, picks up signs of impending danger, it can warn the entire community. Such cooperation is especially important to zebras, since they require water every day.

When the warning system breaks down and a zebra senses its own imminent death, it seems to go into shock and become insensitive to outside stimuli. The adaptation is obviously a protection against pain in animals that are often consumed before they are dead.

Each individual's stripe pattern is slightly different from every other's. Most probably the stripes help to camouflage the animal by breaking up its silhouette. And they may play a part in temperature regulation, since the black stripes absorb more heat than the white ones do.

Black Rhinoceros
Diceros bicornis

Black rhinos seem to have more than their share of special adaptations. Their size alone, up to 2 tons (1.8 m t), is enough to deter enemies, and indeed rhinos have none but man. Their pointed upper lip permits them to browse on vegetation so coarse that it is inedible to most other mammals. And their long horn helps them excavate roots and other foods from the ground. In addition, the tick bird serves the rhino well by consuming its body parasites and by giving out warning signals when danger seems to threaten.

In spite of these advantages, the black rhino is severely handicapped by its limited eyesight and seemingly low intelligence. In fact, the charges the animal allegedly makes on everything from trains to people may be nothing more than frustrated attempts to get a good look at the objects around it. The eyes are situated so far apart that they preclude binocular vision, and the front horn and muzzle area are so located that the rhino must actually try to see around them at times. Stationary objects appear fuzzy or even totally undefined if they are outside a certain range.

Poaching for rhinoceros horns, which supposedly have aphrodisiac powers, still persists in spite of legislation designed to curtail it.

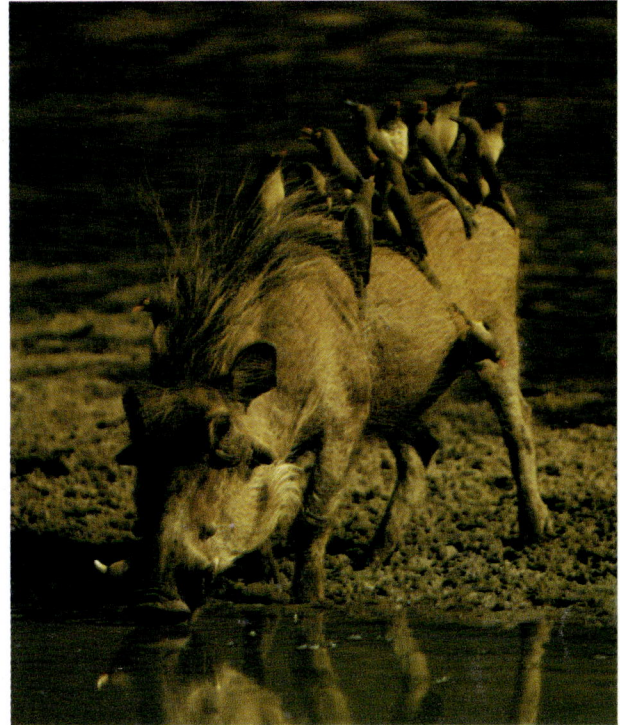

Wart Hog
Phacochoerus aethiopicus

Despite their tusked and warty appearance, these wild pigs are not among Africa's braver species. They live in underground burrows, often the abandoned homes of other animals such as aardvarks. One of their primary means of self-defense is simply to remain inside the burrow. If they must confront a lion or some other enemy, they use their short lower tusks. The larger ones, developed more fully in males than in females, may be just for show, although they do seem to offer the wart hog's face some protection from brambles during feeding. Warts, too, are restricted to the males and serve no clear-cut function except perhaps to make the head appear larger than it really is and to protect the animal's eyes.

Wart hogs feed by day, and small parties of them may be seen trotting across the savanna with their tails held high. The recent proliferation of hippos in certain parts of Africa has helped to encourage the presence of wart hogs, since hippos keep the grass cropped to the wart hogs' favorite length.

Giraffe
Giraffa camelopardalis

The world's tallest animal (up to 18 feet [5.5 m]) has access to high-growing vegetation, certainly; and its prehensile tongue is a help in stripping limbs of their foliage. But the giraffe's unusual height presents definite disadvantages when the animal drinks. Giraffes must spread their legs far apart in order for their long necks to be able to reach down to watering places. (Specialized valves and blood vessels prevent blood from flowing too rapidly into the lowered head.) A running giraffe and a giraffe rising from a resting position demonstrate carefully counterbalanced neck and leg movements.

Giraffes are not thought to have significant sound-producing mechanisms, which is not surprising since inhaled air has such a long way to travel before it reaches a voice box. On the whole, however, sound is not really needed by these animals, since giraffes have few enemies, and their sense of sight is strong enough to keep individuals in touch. Only a very hungry group of lions, for example, will even try to bring down a giraffe (lions kill by strangling, and a giraffe is not an easy victim). Men, however, will not hesitate to kill a giraffe for its tail alone: giraffe tails are made into bracelets, popular tourist items.

Hippopotamus
Hippopotamus amphibius

Hippos spend the nighttime half of their lives eating. The feeding process is efficient, thanks to a lip that crops grass extraordinarily fast. By the time the hippo has digested its food, it has usually submerged itself in water, where it spends most of the day, and where it defecates. While its waste keeps the waterways fertile, it does nothing at all for the grasslands themselves; since most hippos die in the water, their corpses do not help enrich the land either. This one-way system—an animal using the land but giving little or nothing back to it—is among the more unusual ecological phenomena of the savanna.

Hippos in the water are marvelous to watch. They seem designed to loll and float, although they can stay underwater for as long as thirty minutes at a time if they need to protect themselves. When they surface, their tiny ears twirl independently of each other and thus rid themselves of water.

Hippo ivory is much in demand, since it is relatively easy to carve, and ivory poachers are all too common.

African Buffalo
Syncerus caffer

African buffalo are closely related to domestic cattle, and have basically the same simple requirements—sufficient green food and freedom from persecution by man. In recent years they have seldom been granted the latter, at least not for long; instead they have met repeatedly with hunters who covet their long curved horns. Their meat is highly valued also, and buffalo hooves are even used to make ashtrays. It is no wonder that African buffalo have developed a certain wariness. Their reputation for making vicious, unprovoked attacks is, however, undeserved.

These are true herd animals: as many as two thousand may congregate in open country. An old female generally leads the herd. Adult bulls form their own social hierarchies and engage in pushing contests, using their horns. The losers relinquish their females and usually join others like themselves in bachelor herds. These herds are thought to divert the attention of predators from the females and young, and thus to contribute to the proliferation of the species.

In hot weather nothing seems as satisfying to a buffalo as coating its entire body with mud in a wallowing hole. This probably serves to lower body temperature and at the same time to keep insect pests off the animal's hide.

Grant's Gazelle
Gazella granti

Although they sometimes inhabit grassy woodlands, Grant's gazelles thrive in brushy, arid regions, doing well even in midday sun. Their adaptability seems to derive from several different features. First, they can get along without drinking water for extended periods, partially because they often graze at night, when grass absorbs the greatest amount of moisture. Second, their shiny coats reflect heat very effectively. And third, they have an unusual means of cooling the head area. Blood traveling to the brain is cooled as it passes through a series of tiny arteries that run parallel to each other. In hot weather, this adaptation helps to keep the Grant's gazelle's brain several degrees cooler than the rest of its body, thus preventing brain damage.

There are only a dozen or so species of true gazelles, which are a type of antelope. (Antelopes are Old World ruminants of the Bovidae family. In contrast to true oxen, they are generally built for fleet movement and have horns that project up and back.) The Grant's gazelle resembles both the smaller Thompson's gazelle, with which it is often seen, and the impala, perhaps the best-known leaper among African ungulates. Rump and tail markings, horn shape, and face patterns serve to distinguish the three species.

Dik-dik
Madoqua sp.

Dik-diks are antelopes. Yet the adults weigh only about 9 pounds (4 kg) and stand little more than a foot (30.5 cm) high. Their diminutive size works to their advantage by allowing them to get inside thickets to browse, something rhinos, giraffes, and other bulkier animals would find impossible. Even when they browse outside thickets, dik-diks eat very low-growing shrubby vegetation, much of which they grasp with their proboscislike nose. Thus they avoid competition with taller species.

Larger members of the wildlife community are indebted to dik-diks—these fleet and shy ungulates act as sentries by zigzagging through the brush calling "dik, dik" or "zik, zik" at the very first sign of danger. Their own camouflage (dark back, light underparts) is particularly effective. Yet the tiny ungulates often fall prey to hunters, who are well paid for the fur.

Highly territorial, the same pair of dik-diks may occupy a particular piece of land for years. They stake it out with their feces; with secretions from the gland in front of their eyes they mark grasses as a warning to intruders.

* Hartebeest
Alcelaphus buselaphus

Hartebeests, topis, and wildebeests are all classified in the same subfamily, Alcelaphinae. Yet except in their bulk, they are not terribly similar. For one thing, hartebeests are not so totally devoted to a social life style as wildebeests are. Perhaps this is related to their feeding patterns—they seem to survive hard times very well by consuming dried grass. They therefore do not need to be constantly on the move in search of fresh pasturage and so do not have to herd for protection. By the same token, female hartebeests retire to sheltered places to bear their young instead of bearing them in the midst of a traveling herd. The offspring, which are not extremely well developed at birth, are left alone except when the mother returns periodically to suckle them.

Hartebeests' handlebar-shaped horns often make the animals the subject of jokes, as do their rather steeply angled hindquarters. Two subspecies of the species *buselaphus* are endangered, both because of overhunting and habitat degradation or destruction. They are *A. buselaphus tora* and *A. buselaphus swaynei*.

Gerenuk
Litocranius walleri

Gerenuks are complete teetotalers when it comes to water. These lovely gazelles have never been observed to drink in the wild, even where freestanding water is readily available. Such a thoroughgoing adaptation to arid habitat is rare among all but the true desert species, and may possibly help to account for the gerenuk's expanding African range.

This fairly small mammal (about 3.3 feet [1 m] at the shoulder) has evolved the ability to browse while standing on its slender hind legs. This gives it access to leaves and twigs growing as high as 8 feet (2.5 m) above the ground. Acacia leaves are a favorite food, and gerenuks are often seen with their forelegs propped on an acacia's low-growing branches while they eat from higher in the tree. The long neck is able to reach so far up into the branches, in fact, that the animal actually resembles a giraffe when it is feeding. It was no doubt from watching the gerenuk in this position that the people of Somaliland originated the animal's common name, which means "giraffe-necked."

* Sable Antelope
Hippotragus niger

Of all African antelopes, the sable may have the most spectacular horns. Only the scimitar-horned oryx, the addax, and possibly the greater kudu are in its class. It is not surprising that the giant sable antelope, *H. niger variani,* is on the endangered list—its horns are worth a great deal of money to trophy hunters.

Giant sable antelopes were discovered by scientists the most recently of all large African mammals, in 1913. They are a subspecies in the same genus as roan antelopes. The roans are solitary creatures, moving about in search of food either alone or in pairs. Sable antelopes, on the other hand, are herd animals, and may gather in groups of fifty to seventy-five individuals.

Usually shy, the sable antelope will fight lions and other predators with its horns. It often drops to its knees to fight; from this position, too, the male engages in display encounters with other males.

Klipspringer
Oreotragus oreotragus

The stippled-looking antelopes known as klipspringers, or cliffspringers, come by their name naturally: their habitat is limited to areas of rocky outcroppings and cliffs. Gaining a foothold in such terrain would be all but impossible were it not for the klipspringer's hard, rubbery hooves. Only their very tips touch the ground as the klipspringer walks, and when the animal leaps, it manages to land with all four hooves brought together in a very small place.

These animals' coats are brittle, and make a slight rattling sound when ruffled. The coats are also helpful in a rocky environment, since they seem to serve as padding between the animal's body and sharp rocks.

The klipspringer occupies a niche very similar to that of the alpine chamois and the North American mountain goat. Yet it is not closely related to either of those animals; it merely resembles them through having adapted to a similar environment.

Blue Wildebeest
Connochaetes taurinus

Blue wildebeests, also called brindled gnus and white-bearded wildebeests, are known for their herding and breeding patterns, which have been the subject of several scientific studies. Migrating blue wildebeests, stretching single-file across the arid plains, may cover as much as 11 miles (18 km) a day in their search for fresh grass. If they fail to locate tender young blades, they may resort to eating wild melons and other supplementary foods.

These ungulates almost always bear their calves at the start of the rainy season, and 95 percent of the females give birth every year. Perhaps because of this high percentage, blue wildebeest herds can withstand heavy predation—they constitute nearly half the food supply of lions. The wildebeest is probably in much more serious danger at the hands of poachers, who kill them just for their tails, which are made into fly whisks.

Black wildebeests, the blue wildebeest's cousins, are known for their unique displays of ritual clowning. At times they prance about crazily, or so it seems to human onlookers.

Eland
Taurotragus oryx

Africa's largest antelope, the eland is beautifully adapted to arid land. Though it requires up to 6½ gallons (25 l) of moisture a day, it can obtain all of it from the acacia leaves it eats. It therefore does not have to depend on freestanding water or on rich grazing lands. In extreme heat the eland loses relatively little body water, since it pants instead of sweating excessively; and its body temperature rises or falls somewhat, according to the demands of the weather.

With all of these factors in its favor, it is no wonder that the eland is being experimented with as a domestic animal. In Africa the demand for protein is great. If elands are as efficient at converting grass to meat as some people hope, then perhaps they can provide quantities of high-quality protein through their milk (which is also extremely rich in minerals and fat) as well as their meat.

The eland uses its horns to break off high-growing tree branches and bring the leaves within reach. The foliage the eland itself does not consume is then available to other savanna animals. Thus the antelope performs an ecologically important service.

Spiny Anteater, or Echidna
Tachyglossus aculeatus

Spiny anteaters and duck-billed platypuses are the only living representatives of the monotremes, or egg-laying mammals. Their eggs are much more pliant than birds' eggs, and in fact resemble the eggs of reptiles, to which the monotremes are closely related. During the breeding season the female echidna develops a special pouch, which is used to house and nourish the young after they hatch. The females have no nipples, but nurse their young directly from milk glands that open into the pouch. The newborns lap the milk rather than suck it.

Their strong, flattened claws make these animals beautifully adapted for burrowing, and when threatened they dig themselves into the ground until they present only a slight show of spines. If the ground is unsuitable for digging, echidnas will roll themselves into a sphere of formidable appearance. An echidna that has wedged itself among the rocks is nearly impossible to dislodge.

The long snout terminates in a small mouth with a slender, sticky tongue that is ideal for catching insects. The echidna has no teeth, but the rear part of the tongue is serrated for grinding food against the palate. Large salivary glands produce enough mucus to make termites and ants quite palatable.

Red Kangaroo
Macropus rufus

While large, hoofed, grazing mammals were evolving throughout most of the world, kangaroos were developing on the isolated island of Australia. Today the two groups of mammals occupy the same ecological niche.

Kangaroos derive both their speed and their generic name from their long hind limbs—*Macropus* means "big foot." The hind legs propel the animals forward in leaps of 6 to 12 feet (2–3.6 m). The strong tail functions as a counterbalance during leaps, as a prop during feeding, and even as a body support during fights, since the kangaroo defends itself by slashing with its foreclaws while its body is in an upright position. By comparison with the massive rear appendages, the forefeet are delicate indeed, and are used primarily in walking.

The kangaroos and their close relatives the wallaroos are the largest living marsupials. No doubt they are also the most well known, especially for their method of reproduction. The single young that the female bears is less than an inch (2.5 cm) long and only partially developed at birth. It finds its way into the mother's pouch, where it stays for about six months as it develops. The young kangaroo, known as a joey, then leaves the pouch, but continues to nurse for several more months.

A form of delayed implantation probably assures the female of maximum reproduction during favorable climatic conditions. She mates again soon after her first joey is born, but does not actually bear the second young until the first has left the pouch. Drought or other natural calamities will usually interrupt the process by claiming a high percentage of the fetuses and newborns. For example, a newborn will perish if the mother finds nursing an intolerable strain when fresh vegetation is unavailable.

Hairy-nosed Wombat
Lasiorhinus sp.

Soft pelage, pointed ears, and a hairy muzzle distinguish wombats of the genus *Lasiorhinus* from those of the other genus, *Vombatus*. The members of both genera resemble small bears in their thick bodies and short legs. They resemble rodents in their constantly growing incisors and in their manner of moving their jaws when they chew vegetation. They are marsupials, not rodents, but they fill the same niche that several rodents fill in other parts of the world. While they owe their excellent burrowing ability more to their claws than to their teeth, they can chew right through roots and other obstacles. They have been known to make burrows of up to 100 feet (30.5 m) in length, although this is probably unusual. They are the only large Australian marsupials that burrow at all.

Tunneling plays an important part in the wombat's ability to conserve moisture, which species in arid areas do with admirable efficiency. Wombats stay underground during the day, for example, where the humidity is relatively high; and the *Lasiorhinus* species do not seem to sweat. Furthermore, wombats expel very dense feces, and so lose little moisture through their solid wastes.

The range of at least one *Lasiorhinus* species is diminishing very fast, as a result of the animal's interference with man's agricultural activities.

Rufous Rat-Kangaroo
Aepyprymnus rufescens

Rat-kangaroos, diminutive members of the kangaroo family, differ from their larger cousins in basic dentition. Their canines are relatively large and indicate their generalized diet of roots and tubers, and even their ancestors' insectivorous habits. The true kangaroos, on the other hand, have very small canines if any, evidence of their grazing habits.

The rufous rat-kangaroo is one of the largest rat-kangaroos, weighing 4½ to 6½ pounds (2–3 kg). By day it nests aboveground in the grasses, and like a few other small mammals, it transports bundles of nesting materials to a chosen site by wrapping its prehensile tail around them. Extremely agile, it will usually wait until intruders are very near before it dashes to shelter, preferably in a hollow log. It seems to have very little fear of people, and in fact has entertained many a camper with its lively nocturnal activities. Foxes take quite a few of these rat-kangaroos, and are responsible for their declining numbers.

Ghost Bat
Macroderma gigas

Ghost bats, so called for their pale coloring, are voracious eaters. Their carnivorous and cannibalistic habits include the frequent consumption of frogs, grasshoppers, birds (including at least one species of pigeon), and several smaller insectivorous bats. Smaller bats sharing cave space with this species, in fact, have been found wedged deeply into rock fissures, perhaps in self-defense. Ghost bats swoop down on their prey and engulf it in the membranous skin connecting their limbs. They kill it by biting it in the head area, and they eat the entire body—bones, teeth, and all. Their own large size (head and body together measure about 5 inches [125 mm]) and 2-foot (61-cm) wingspan make such a life style feasible.

This is the only Australian representative of the Megadermatidae, a family of leaf-nosed bats with members in Africa, Asia, the Philippines, and part of Indonesia. It is distinguished from other leaf-nosed bats by the asymmetrical end of the ear tragus, a projection inside the outer ear that is thought to determine the direction from which echoes bounce back to bats.

Despite their size, ghost bats are so furtive that they are often not noticed by people in the mines and caves that shelter them by day.

Spectacled Hare-Wallaby
Lagorchestes conspicillatus

Hare-wallabies are so named because of their close resemblance to European lagomorphs. Like hares, these marsupials rest in aboveground "forms," or depressed areas in the soil. When startled, they hop away at the last possible minute, with great speed.

Weighing only about 4½ to 6½ pounds (2–3 kg), hare-wallabies are the smallest of all the true kangaroos. The popular names for these marsupials—the various wallabies, wallaroos, and kangaroos—have been assigned on the basis of the length of each species' hind foot. In the hare-wallaby this measurement is usually less than 6 inches (15 cm). Yet even at this small size, the hind foot is enormous in proportion to the rest of the body, and is *the* adaptation that makes the hare-wallaby's grazing, hopping life style possible.

All five hare-wallaby species have distinctive eye markings, and in *L. conspicillatus* these markings are bright orange. All five are becoming rare on the Australian mainland, *L. conspicillatus* being the only one still found in considerable numbers. The reason for this decline probably lies in the alteration of habitat by domestic sheep and cattle, which trample native grasses.

Long-haired Rat
Rattus villosissimus

Australian rodents known as bush-rats are interesting because they are so similar to one another as to be almost indistinguishable to the layman. This similarity seems to indicate how recently these rats have become specialized for various ecological niches, especially compared with such widely divergent mammals as the various possums, kangaroos, and others that became specialized long ago.

The long-haired rat, which exhibits a buff coloration that blends with its savanna habitat, undergoes population fluctuations not too different from the lemmings'. Every five to seven years the numbers of long-haired rats seem to multiply beyond belief, and the rodents literally swarm across the savanna. They consume enormous quantities of vegetation, and they leave behind them a wake of ravaged stems and seeds. Even the roots of large trees are damaged, and cannibalism among the rodents themselves is rampant. Eventually, however, the populations are reduced to their original stable proportions, and individual long-haired rats may once again be difficult to find.

Nine-banded Armadillo
Dasypus novemcinctus

Its means of self-defense is this mammal's most distinctive characteristic. Even its common name, armadillo, is derived from a Spanish word that means "armored" and refers to the plates of leathery skin that the animal bears. Its muscles are so strong that the armadillo can curl its 2-foot-long (61-cm) body into an impenetrable ball almost instantly. Unlike the three-banded armadillo, however, it cannot draw its head into its curled-up body. Therefore the head remains vulnerable. Armadillos caught in the act of trying to escape by burrowing (another defense tactic) may wedge their plates into the soil, thereby making themselves immovable. If allowed to continue burrowing, they can dig themselves completely out of sight in a matter of minutes.

Nine-banded armadillos have such poor eyesight that they are nearly blind, but this handicap is compensated for by their keen sense of smell. They are said to be able to sniff out worms and insects some 8 inches (20 cm) deep in the ground. Feeding is usually a nighttime activity and is accompanied by almost constant grunting.

There are actually five species of nine-banded armadillos. *D. novemcinctus* has greatly increased its range within the past century; it now covers nearly all of South America and quite a bit of southern North America.

Maned Wolf
Chrysocyon brachyurus

The maned wolf, with its long legs and shaggy coat, is somewhat bizarre in appearance. Its rear legs are a bit longer than its forelimbs, giving the animal a slightly asymmetrical look. They also make going downhill quite difficult. The stiltlike legs are valuable, however, when the canid goes uphill or maneuvers among the tall grasses that cover portions of South America. And they put the animal's eyes high above the level of vegetation.

The maned wolf sometimes uses its teeth to excavate rodents from the soil. It also eats quantities of roots, as well as many birds' and turtles' eggs, snails, sugar cane, and several types of fruits. Such a large and varied diet of ground food qualifies the maned wolf as a terrestrial omnivore.

This canid hardly fits the popular stereotype of the wolf—its temperament, far from being aggressive, is almost shy. It does not live in a highly structured social group typical of the wolves and many other wild dogs, but instead leads a solitary life.

Entellus Langur
Presbytis entellus

Entellus langurs are the sacred monkeys of India, and as such are afforded a unique place in Indian society. They are not harmed, or even scolded, regardless of the amount of food they take from people, nor are they excluded from city streets, dinner tables, or temples. They actually consume very little "human" food, however, since their diet is restricted to flowers, tree fruits, and particularly leaves. (Langurs are leaf-eaters and have multi-chambered stomachs, not unlike a ruminant's, for breaking down cellulose.) A great deal of time is required for the digestion of leaves, just as it is for the digestion of grass and cud in the ruminants, and langurs may be seen sitting quietly in the trees for much of the day, digesting their food. Perhaps the rather pensive and even prayerful attitudes that they adopt at these times led to their sacred status. They derive additional status from a legend in which langurs helped to rescue a revered member of Indian royalty.

Four-horned Antelope
Tetracerus quadricornis

The only member of the Bovidae family to have four horns is this dainty antelope, which weighs no more than about 44 pounds (20 kg). Sometimes its front horns are nothing but slight rises on the forehead, and the back horns, while considerably larger, measure only about 4 inches (10 cm) at most. Despite the small size of the horns, the animal's head is considered a trophy.

Four-horned antelopes are much shyer than other Indian antelopes. They inhabit lightly treed areas as well as open spaces, and they need to drink regularly. They prefer to remain solitary or in the company of one other individual, rather than gathering in herds. When threatened, four-horned antelopes may sneak away, if possible, or duck down until the danger passes. Fortunately, they can move quickly, if somewhat jerkily, when they need to.

Chital, or Axis Deer
Axis axis

While most deer living in temperate climates experience definite rutting seasons, the prolific chital of India and Ceylon breeds year-round, though most of the fawns seem to be born during the cooler months of fall. The only time that the sexes even separate is when the stags isolate themselves for a short while to shed their antlers.

Transporting populations of chitals to new lands seldom seriously disrupts the animals' breeding habits. The introduction of axis deer into New Zealand, in fact, resulted in a veritable population explosion. Professional hunters were finally called in to kill considerable numbers of chitals in order to protect the island's native flora.

Axis deer are found in large herds, sometimes numbering as many as one hundred head. They are diurnal grazers and browsers, and may often be seen in the company of nilgais, blackbucks, and various other species.

Blackbuck
Antilope cervicapra

Blackbucks have evolved a fascinating system of defense and retreat. The animals live in the open, in herds of at least a dozen or so members. When notice of danger spreads, each blackbuck leaps high into the air until the entire herd is bounding across the plains. They move this way for several yards, and then they run—at such a fast pace that even greyhounds are hard pressed to catch them.

The male blackbuck's threat display includes showing off the white coloration on its rump and bobbing its marvelously horned head. Such postures are important for the establishment of territories, and dominant blackbuck males are highly possessive of their harems, at least during rutting season. (Young males group together in separate bachelor herds until they, too, are old enough to attract harems.) After rut, territorial boundaries lose their importance altogether, and several herds intermingle until the next rut.

Blackbucks evolved from a forest species that was forced to adapt to the grasslands as India cleared more and more of her land for cultivation.

5

The Temperate Grasslands

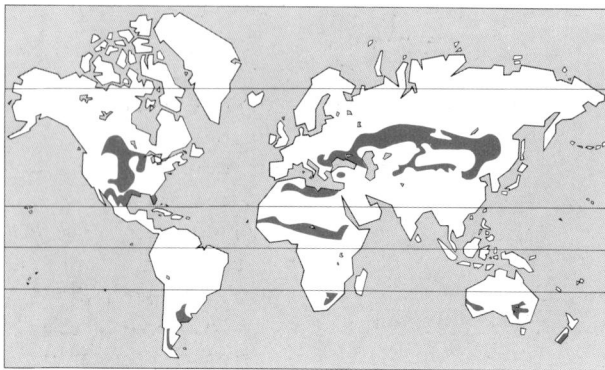

Endless wiry grasses and the bright blooms of sunflowers and wild thistles—these are the temperate grasslands. They occupy the middle latitudes in the northern hemisphere, the lower latitudes in the southern; and they have many local names. In Eurasia they are known as the steppes; in North America, as the prairie or plains; and in South America, as the pampas. Though they are not covered here specifically, limited temperate grasslands occur in southern Africa and Australia as well. Thus, this biome is found on every continent in the world.

The character of the rainfall is the significant force here. The amount (10–30 inches [25–76 cm] per year) is not extremely low, and Argentina's pampas receives slightly more than the average elsewhere, about 35 inches (89 cm) per year. But this moisture is precious because of the sporadic way in which it falls. In late spring and early summer, cloudbursts darken the sky at unpredictable times, then may not occur again for months. Drought has been known to persist for the greater part of a year, and the rate of evaporation is extremely high.

Temperatures also vary widely here from season to season. In the northern hemisphere temperatures may shoot to 90° F (32° C) or higher in the summer, and may fall to 5° F (-15° C) or lower in the winter. In the southern hemisphere the range is not so wide, since maritime influences are stronger in the more southerly, narrower parts of South America and Africa.

Soil in this biome is as rich as any on earth. The type known as chernozem is blessed with a top layer 1 to 2 feet (30.5–61 cm) deep, the bulk of which is fibrous material such as the absorbent interlaced roots of grasses. Unlike some forest soils, chernozem does not lose its nutrient compounds to the lower layers of ground. Instead minerals stay near the surface, where grazing animals can ingest them in their food.

Because the soil and its vegetation can support them, domestic grazing herbivores (sheep and cattle, mostly) are now widespread in this biome. Grazing and grain-growing, in fact, have erased all but a few specially preserved stands of the original North American prairie. And Argentina, long known for its beef cattle, replaced its native grasses decades ago with a man-made ecosystem. In Eurasia, too, the face of the grasslands has been changed so drastically that today the vegetation and animal life bear little resemblance to the area's original species. No doubt the changes were inevitable, for they meant food for millions of people. But one can hardly help bemoaning the passing of the pronghorn, the wild horse of the steppes, and the other grassland mammals that have been replaced by wheat, cows, and machinery. And one must remind oneself that the temperate grasslands of five centuries ago, and even some of the species mentioned on these pages, are now museum pieces at best.

Burning grasslands is a common and very old practice in many cultures. Originally it had the obvious advantage of flushing game animals into the open. It also improved the quality of pasture land by leaving ashes that absorbed the sun's heat readily, thus warming the earth. In the early spring, this warming speeded the melting of the snow, and gave the pasture plants a head start on growth. Though burning is still important for these and other reasons, the practice has been heavily questioned in recent years, and in more developed cultures it is rapidly becoming a thing of the past.

Many scientists believe that fires, either man-made or natural (lightning fires), are in some way responsible for the evolution of the grasslands. Fire does help keep down

Preceding pages:
Herd of grazing bison, midwestern
United States

tree growth by killing seedlings, and the absence of trees is a prerequisite for the growth of grasses. The few trees that manage to survive in this biome are usually found along streams or riverbanks.

With few or no trees to break up the landscape, the sky here seems enormous, and the flat or rolling land a monotonous eternity. On closer inspection, however, one finds the grasses themselves to be fascinating. Most of them are easily distinguished from other types of plants by the parallel veins in their leaves. To survive in this arid environment plants must be stiff and coarse, and these characteristics typify the grasses of this biome. Some species have evolved leaves sufficiently tough or thorny to be unpalatable to herbivores. Others are heavily grazed, yet because they grow from the base, clipping by animal teeth destroys only their current foliage, and not the entire plant. Some plants have even evolved a slow rate of growth in response to heavy grazing.

Many grasses of the temperate zone propagate themselves via underground shoots called rhizomes or surface runners called stolons. This, too, helps to protect against overgrazing. And it helps the grasses to survive drought as well, since the buds on the subterranean shoots remain alive even when the aboveground parts of the plant dry up. When moisture finally does become available, the buds begin to sprout. And the new plants that they produce may eventually form a sod so dense that competing species are thoroughly discouraged from growing.

Because the leaves of grasses grow nearly vertically, each leaf gets its own light requirements without blocking light from the other leaf surfaces. Photosynthesis proceeds easily, thanks to this arrangement. During extended drought many grasses simply stop growing; and even on hot, windy days during times of ample rainfall, some grass leaves minimize evaporation by rolling their edges toward the center and thus reducing the amount of surface area exposed to the air.

Most of the grasses that grow in the middle latitudes are necessarily perennials, the growing season being too short to accommodate plants that "start from scratch" each year. Where chewing, rubbing, digging, pawing, or some other type of animal action has stripped the ground of vegetation, however, annuals often take hold. They act as a temporary soil cover until the perennials have had a chance to reestablish themselves and develop their often elaborate root systems. Ecologists can sometimes discern a grassland's past weather conditions or animal activities by noting how plentiful or sparse the perennials are. This is so because extreme and prolonged drought has the same result as overuse by animals—a predominance of annual weeds where perennial grasses more typically grow.

As aridity increases in the grasslands, plants become more scattered, and they become much shorter. In the North American grasslands aridity increases as one moves from east to west. Within this zone three types of grasslands can be identified, based on the height of the grasses. In the easternmost part of the prairie tall grasses such as the big bluestem are the dominant native form. They often attain a height of 5 to 8 feet (1.5–2.4 m). Although some of the tall grasses form sod while others grow in bunches, the look of the tall-grass prairie is that of a continuous cover of vegetation. The mixed prairie, composed of grasses 2 to 4 feet (61 cm–1.2 m) high, is found farther west. "Mixed" is an appropriate name for this type of grassland, since it comprises both mid-height and dwarf-sized grasses. Still farther west the short-grass prairie begins. It is composed of 6-inch to 1½-foot-high (15–46-cm) bunch grasses that grow very sporadically until they either phase into desert shrub and sagebrush or meet the mountains.

In Eurasia the steppes become more arid as one moves from north to south. The meadow steppe bears the tallest grasses, although their maximum height rarely exceeds 4 feet (1.2 m). Many of them grow closely with the flowering herbs that give the meadow steppe its lovely warm-weather blooms. This type of steppe gives way to a type of vegetation roughly similar to that found in the North American mixed prairie. And the mixed-steppe vegetation, in turn, gradually leads into a small area of fescue and other dwarf grasses that make up what is known as the dry steppe of Eurasia.

In South America, too, the grasslands may be subdivided, as one moves southwest, into the humid pampa, the dry pampa, and the scrub vegetation known as monte, with some grasslands even occurring on the mountain plateaus of Bolivia and Peru.

Grasslands and herbivores are natural complements. The smaller herbivores, most of them colonial rodents, consume enormous quantities of grass seeds and foliage annually, and many use the stems and leaves of grasses to build their homes. Burrowing occupies much of the rodents' time and is facilitated by the animals' teeth and limbs, which are often specialized for excavating soil.

Burrowing not only provides rodents with housing and an easy avenue of escape from predators; it also benefits the grasslands considerably by keeping the soil well mixed and therefore suitable for diversified vegetation. For instance, one species of marmot is capable of excavating subterranean passageways that total some 200 feet (61 m) in length. Such dedicated miners bring soil lying deep within the ground up to the surface at a rate of 3 to 6 cubic feet per year. This one churning process alone makes calcium and other compounds available to seeds that might not germinate without them. Conversely, rodents often take organic matter down into the soil, to be used as nesting material. Thus they provide the soil with new elements even at a subterranean level. Burrowing also keeps the soil texture varied, as evidenced by both the hard-packed earth and the loose soil that often surround different burrow entrances.

Erosion and rodents are closely linked in the grasslands, the former rarely occurring where the latter have been at work. Some gophers, for example, are said to favor the subterranean parts of certain thistles and shrubs with long tap roots, leaving the grass roots alone; and it is grass roots that hold water rather than letting it run off. Thus the simple process of a rodent's food selection

leads to the prevention of erosion.

Rodents, of course, are generally prolific mammals, and without the natural checks of predation, can reach plague proportions in the grasslands. Unfortunately, human beings contribute to this imbalance each time they kill a predator by poisoning, shooting, or some other ill-conceived scheme.

Large herbivores, ungulates to a great extent, attain their greatest numbers in the temperate and tropical grasslands. In fact, many scientists believe that large herbivorous ungulates necessarily evolved after the grasslands did. It is true that forests cannot support large numbers of grazing animals, not only because forest plants cannot tolerate constant grazing, but also because large herbivores themselves are not adapted to eat primarily broad-leafed herbs and green foliage. Many of these herbivores are adapted, by means of their ruminating digestive systems, to "chew the cud," or graze large amounts

Tall grasses in a marshy area of the Argentine pampas

pronghorns than they are a single individual. Herding also helps in the propagation of species, especially in those ungulates that inhabit the more arid grasslands. Aridity dictates an almost continual movement in search of food. Were the nomadic animals to travel singly, they would most certainly meet less often during mating season.

Speed is also a characteristic of the grassland ungulate. When threatened, its only means of escape is to run from its attacker. And although coyotes may wear out a pronghorn by continually circling and dogging it, they will rarely if ever simply outrun one.

Vision is strong in most grassland species. Lack of cover gives predators an advantage, but good eyesight on the part of the prey provides ample warning. Or, where cover is too high for some of the smaller animals to see above it, specialized locomotion and good vision work together. This is true of certain lagomorphs, whose hopping puts them above the height of grasses long enough to take a good look around. Yet by the time a predator has pounced on the spot where a hare appeared, the hare is far away.

Most grassland carnivores are not really limited to this biome, but occur in forests as well. Because they are at the top of the food chain, they are not subject to the same type of predation pressures as the lower-ranking mammals. Yet they are no less susceptible to parasites and other maladies than the prey species are, particularly when they are not in perfect balance with their surroundings. When they are too plentiful and consequently very competitive, or too undernourished because of a decline in prey and consequently not resistant to disease, they too succumb to the checks of nature.

With the exception of the bison, temperate-grassland mammals are usually not seen by observers for a great part of the year. Many, such as the pronghorn, escape the cold or the heat by retiring to nearby woods. Others hibernate or estivate. Whatever method they use, these mammals are adapted to take advantage of the livable weather and retreat from the climatic extremes that characterize their grassland home.

of grasses which pass through two separate stages of processing before they are broken down for utilization in the body. Grassland ruminants include deer and antelopes, the pronghorn, and the bison.

The need for this complex type of digestion is obvious when one considers the nature of grasses. Many are coarse in texture, and all are relatively low in food value. Because of this latter characteristic, an animal as large as, say, a bison must consume great quantities of food in order to obtain adequate nourishment. Rumination allows the animal to crop rather quickly the large amounts of food it needs, then retreat to a sheltered place in which to digest it slowly. Even so, grazing occupies much of an ungulate's time in the temperate grasslands.

It is no accident that grazers are also herd animals. This adaptation helps ensure the safety of a species in vast, open spaces without cover. Even a pack of carnivores such as coyotes are far less likely to attack a dozen

Steppe Lemming
Lagurus lagurus

Two species of steppe lemmings are adapted to arid habitats in the Old World grasslands, and a third species inhabits the New World plains. They differ from many other rodents in that they eat quantities of green vegetation, though some also store hay for winter use. Only one out of every ten or so *L. lagurus* individuals harvests hay for an entire colony. When temperatures fall below 5° F (-15° C), this species remains in its burrow with the entrance plugged.

Like all rodents, steppe lemmings have two upper and two lower incisors that grow continuously from the base. These teeth would lose their usefulness altogether if gnawing did not keep them in good condition and at the proper size. The inner surface of the incisor is much softer than the outer. Thus, gnawing wears away the inside, leaving the outer surface sharp and hard.

Lemmings of this genus are perfectly equipped for burrowing, with their small ears, abbreviated tail, powerful claws, and hairy soles that are no doubt useful in providing traction.

Spotted Souslik
Citellus suslicus

Old World species of the genus *Citellus* are known as sousliks, New World species as ground squirrels. Studies on some individuals in the latter group have shown them to be true hibernators that enter their sleep well fed. Prior to hibernation, in fact, satiety centers in the brain apparently function poorly if at all, allowing the ground squirrels to eat considerably more than they can during other seasons. Food is converted into fatty tissue known as brown fat. In extremely cold weather the brain's heat-regulating mechanisms trigger the burning of this fat, and the hibernating animal is warmed in a matter of a few hours. (The brain maintains a constant temperature even when the animal's body-surface temperature drops drastically.)

Citellus suslicus is a true steppe-dweller, yet it has accommodated itself to cultivated areas by constructing its burrows only in areas that are not directly beneath plowed ground, where farm implements could destroy them. It often excavates its complex burrows in the dry ground beneath a stone next to a grain field, or on a well-drained slope. Although it does not need to store winter food, it frequently makes small stores of grain, bulbs, and other vegetable foods, perhaps for inclement days when it stays in the burrow. On warm, sunny days it feeds outside, where its whistling can be clearly heard.

Common Hamster
Cricetus cricetus

Golden hamsters are popular pets with an ancestry in the Old World, where about fourteen species of hamsters are found. In dry terrain, and particularly on the steppes and the edges of the desert, hamsters subsist during the harsh winters on stored supplies of cereal grains and other foods. They interrupt their long winter sleep to eat, having transported food to their burrows in their large cheek pouches. (When filled, the pouches give the animals a rather grotesque appearance.) Sometimes hamsters inflate their cheek pouches before swimming, giving themselves extra buoyancy.

Scrupulously clean mammals, hamsters build latrines into their burrow systems. If for no other reason than this, they make good pets.

The common hamster is unusual in its coloration, having a black stomach with lighter areas on the sides and back. All members of the genus fill an important ecological niche by providing food for members of the weasel clan, birds of prey, and other predators. However, they can be definite pests in agricultural areas.

Mole-Vole
Ellobius talpinus

The presence of this mammal can be detected by the mounds of soil the mole-vole ejects during digging. That is usually all the evidence of itself it provides, however, being extremely shy and nervous when it comes aboveground in daylight. Its reluctance to surface is no doubt a means of self-defense, since the little rodents are often gobbled up by birds of prey and other enemies as soon as they become obvious on the steppe. At night, however, they venture on extended excursions, as much as 500 yards (457.5 m) from home.

This rodent uses its incisors to excavate two types of tunnels—a system of feeding tunnels just under the surface and a nesting tunnel farther down. The former may meander for quite some distance, and the mole-vole consumes roots and bulbs as it bores along. Small stores of roots are often placed near the nesting chamber as well.

Ellobius has a wide distribution—wherever soil conditions permit its tunneling in the steppes, semideserts, and even deserts of Eurasia, and to altitudes of 13,000 feet (4,000 m).

Birch Mouse
Sicista subtilis

The birch mouse is extremely agile and climbs even the smallest of twigs, which it grasps with its toes. It feasts on berries and other vegetable matter, which it seems to require only at intervals: this mouse can go without eating longer than other mice can, and compared with the shrew it is a skimpy eater at best. When it does find large supplies of food, however, it consumes them all at once.

Even the jumping mice leap no better than this species, and its long tail seems to function both as a balancing organ and as a support during climbing. Like the tail of the harvest mouse, it curls readily around stems and other objects.

Birch mice cannot tolerate exposure to the cold—they become chilled and die quickly. Since they cannot afford to be aboveground in winter, they den up in their burrows for about six months. *S. subtilis* is considered a true steppe inhabitant, yet it may range into semidesert and hilly areas, where it seeks out cultivated grain fields.

Marbled Polecat
Vormela peregusna

The marbled polecat's coloration is perhaps its most distinctive feature. The mottled back and the dramatically masked face stand out against the animal's long bristly tail when it is curved over the back. The total effect is bold enough to make many animals retreat, especially when this mustelid bares its teeth and gives off its fetid odor as well.

The chernozem belts of Russia provide ideal habitats, since the marbled polecat excavates burrows deep within the soil. Its food consists of small rodents and hares, lizards, frogs, and birds. Most hunting is done at night, but marbled polecats have also been observed sunbathing at midday. Though the pelts of this mammal are not extremely valuable, they are collected in some areas. Marbled polecats are similar in form to *Mustela putorius*, another polecat of Asia, Europe, and Africa.

Saiga Antelope
Saiga tatarica

An astonishing reproductive rate is responsible for this species' survival. The females mate when only seven or eight months old and bear offspring soon after their first birthday. (Young are usually born in the spring.) Well over half the time saiga antelopes produce twins, which is rare among ungulates. The males assemble harems of five to forty-five females each. Males who lose the autumn rutting battles are usually too weak to survive the steppe's severe winter and simply starve to death. This is a harsh but effective means of assuring that only the strongest individuals reproduce. Were their propagation patterns not so effective, saiga antelopes would certainly have died out in the early nineteenth century, before hunting them for their horns (believed to have medicinal value) was banned. Today an estimated three million saigas roam the Russian steppe, making them the most numerous wild ungulate in the Soviet Union.

The downward-pointing nostrils of this northern antelope prevent blowing sand and snow from entering and congesting the nasal passages. They also contribute to the animal's bizarre appearance.

Black-tailed Prairie Dog
Cynomys ludovicianus

Prairie dogs are colonists. Their "towns" are subdivided into smaller units of between eight and ten individuals, all of which seem to be social equals. Much greeting, grooming, sunbathing, and playing characterizes the activities of these diurnal mammals. When danger threatens, they communicate via sharp barks, yaps, and whistles. Their highly organized societies undoubtedly have significant survival value, since members assume responsibility for each other's safety in the huge expanse of the prairie. Burrow mounds provide important lookouts and also prevent rainwater runoff from flooding the tunnels. The size of the mounds is important to prairie dogs, and they will excavate or kick up enough soil to make a mound of at least a foot (30.5 cm) in height.

Unfortunately, though the black-tailed prairie dog was once found in large numbers throughout its range (much of the plains of the United States and into Mexico and Canada), today it proliferates primarily in national parks. It has been persecuted—shot, trapped, and poisoned—by ranchers whose cattle were placed in direct competition with it for grass. The black-footed ferret, a chief predator of this prairie dog, has suffered along with it and is now one of the rarest native North American mammals.

American Harvest Mouse
Reithrodontomys sp.

Plains mice and western harvest mice are tiny rodents that are often confused with house mice, *Mus musculus*. In fact, they seem to thrive in the company of house mice and deer mice, and even form social groups with those species. And they use the grassy runways of field voles as their own.

Harvest-mouse nests are often constructed aboveground, usually in short grasses or shrubs. They are spherical, about 3 inches (7.6 cm) in diameter, and lined with the down of cattails and other plants. Populations of harvest mice tend to fluctuate quite a bit and may diminish from about twelve individuals per acre to only one per acre over the course of a winter. Most probably breeding stops altogether during severely cold weather, then resumes in the spring.

Active year-round, members of this genus bend the stems of grasses in order to reach the seeds on top, and they also eat green vegetation and insects. The tiny nocturnal mammals are not known to live more than a year.

Pocket Gopher
Geomys sp.

The pocket gopher is named for its large, fur-lined cheek pouches, which open on the outside of the rodent's head and not into its mouth as squirrels' cheek pouches do. These satchels transport vegetable food to storage areas within the burrow, where the gopher empties them by turning them completely inside-out. A specially adapted muscle returns them to their usual position.

This small mammal's incessant tunneling aerates grassland soils but also causes considerable damage to crop lands. Gophers even chew at such materials as the metal casing on underground cables if they happen to be blocking a pathway. Such gnawing presents no health hazards, since the gopher's lips close behind its incisors during burrowing and prevent inedible material from entering the mouth.

In winter pocket gophers line their passageways in snow with soil. When the snow melts, the linings may be seen lying toppled about in the countryside before the rains have had a chance to wash them away.

Prairie Vole
Microtus ochrogaster

Prairie voles are both surface feeders and burrowers, depending on the season. In summer the vole darts about under a canopy of vegetation, clipping alfalfa and other grains, grasses, and clovers. In autumn, in addition to excavating their own subterranean chambers, the voles may utilize mole or pocket-gopher tunnels, which are usually only 3 or 4 inches (7.6–10 cm) underground. Winter, however, sends prairie voles as far as 2½ feet (76 cm) down, where they inhabit cozy globe-shaped nests of shredded vegetation. There they draw on stores of tubers, bulbs, and roots.

Relationships among individual prairie voles seem fairly peaceable. When quarrels do erupt, they are often between females, which have strong protective instincts toward their young. (They are capable of bearing a litter about every twenty-one days.) However, mothers will sometimes eat their newborns, and predators are plentiful on the prairie. Both of these factors help regulate vole populations.

Coyote
Canis latrans

Coyotes are adapted to resist their greatest threat on the prairie—man. More poisoning and shooting programs have been aimed toward eradicating this animal than perhaps any other, but the coyote has survived. The reason is simply that coyotes are smart, and have learned, through being persecuted, to outwit most enemies.

Hatred of these wild canids is widespread primarily because they prey on domestic sheep, turkeys, and calves. But coyotes will eat nearly anything, including carrion, and are an important check on rodents. They help to keep the populations of larger wild animals strong by culling the infirm from the herds.

Portraits of the coyote often depict the animal with its head back and its mouth open in a howl. The sound, which people have come to associate with the American West, is among the most mesmerising music in the animal kingdom.

Spotted Skunk
Spilogale putorius

Great horned owls relish this mustelid but may receive the special ''skunk treatment'' of a handstand and a spray. This skunk's scent glands are located beneath the tail. In defensive postures, therefore, the bushy white-tipped tail is flipped over the animal's back in preparation for ejection of the spray.

Spotted skunks seem to throw their scent more often than striped skunks do, especially when they are ensnared in traps. And they may prance about on their hind legs simply for the fun of it. Altogether, they are much livelier, prettier, less sluggish mammals than the larger striped skunks. They occupy various habitats throughout most of North and Central America, where they may climb trees in search of small animals and fruit. And they share a behavioral characteristic with the mongoose—they throw backward between their legs various eggs that they find.

Because it is known to transmit rabies, this mustelid is often called the hydrophobia cat or phoby cat, and it also goes by the name civet cat, though it is not a feline at all. Perhaps the daintiness of the western species, *S. gracilis*, is partially responsible for these names.

▲ *Coyote* ▼ *Spotted skunk*

* Pronghorn
Antilocapra americana

Pronghorns are the fastest mammals in the New World; they have been known to outrun even speeding automobiles. Yet they are so curious that their speed does not always guarantee their safety. As often as not a pronghorn will investigate a situation from which it really should have run.

Like bison, pronghorns suffered severe reductions in their numbers in the last century, but are now protected from slaughter. Their herd instincts are strong and their method of alerting each other to danger is a lovely thing to watch: they flash their white rumps in warning. Newborns are protected from coyotes and other predators not only through this system, but also by being kept very flat on the ground until they are strong enough to travel. That way they are difficult to spot, even on the open prairie.

This is the only surviving member of the Antilocapridae family, which flourished during the Miocene and Pliocene epochs. Two subspecies, *A. americana peninsularis* and *A. americana Sonoriens*, are endangered because of overhunting, habitat destruction, and competition with domestic species for food and water.

Bison, or Buffalo
Bison bison

Bison are hard on their environment. Herds wander almost constantly in search of fresh grazing lands (bison are ruminants), and can rid an area of vegetation in a very short time simply by wallowing in the dust (this helps keep body parasites to a minimum).

It seems inconceivable that these animals, the largest North American land mammals, could have been the victims of the wholesale slaughter that nearly wiped them out in the nineteenth century. Yet millions of them were hunted for sport and other reasons, and were often left to die on the plains without even being touched for food or leather. At one point their population totaled only about five hundred individuals. Little by little they have recovered, and today they are found in rather plentiful supply in private and public herds.

The only other member of the genus *Bison* is the European wisent (*B. bonasus*), a species that inhabits woodlands. It has a history of exploitation very similar to the buffalo's.

▲ Plains viscacha ▼ Tuco-tuco

Cavy, or Guinea Pig
Cavia sp.

Voles and lemmings of the northern hemisphere occupy the same ecological niche that guinea pigs of South America do—that of a surface grazer. In parts of Uruguay guinea pigs are so plentiful that they have actually ravaged the grasslands. With its small ears and compact body, it is not surprising that the guinea pig can move through its mini-world of grass stems with ease.

Many authorities consider the guinea pig the first rodent ever to be domesticated. *C. porcellus*, the domesticated form, is eaten in parts of South America today, and has been eaten there since at least the sixteenth century. Early Spanish explorers even took guinea pigs home with them from South America, and they soon became a common delicacy on dinner tables all over Europe. Today the animal is widely valued as a pet, since it has very little odor and is relatively docile. It is also one of the most popular species used in medical research, particularly research dealing with heredity, toxins, and food values.

Plains Viscacha
Lagostomus maximus

Colonies of plains viscachas are not as plentiful as they once were. However, they may still be seen across the Argentine pampas; their presence is indicated by meticulously cleared land dotted with piles of clipped vegetation and other debris around the moundlike openings to their burrows. It may be that these members of the chinchilla family clear land to assure visibility and access, since viscachas can run up to 25 miles per hour (40 kph) and make rabbitlike leaps and zigzags when pursued.

Plains viscachas fill the same herbivorous niche in South America that marmots do in many other parts of the world. As far as anyone knows, this is the only rodent that exhibits a visible secondary sex characteristic in addition to size difference. Male viscachas have noticeably long and rugged mustaches, the purpose of which is not understood.

Tuco-tuco
Ctenomys sp.

Tuco-tucos' burrows are fairly sophisticated structures. These diurnal mammals loosen pampas soil with their incisors, then sweep it away with their hind feet. Tunnels usually dip less than 1½ feet (46 cm) below the surface of the ground and include a grass-lined nesting area as well as a food-storage room or two. Grasses (which also constitute part of the tuco-tuco's diet) may be sized and positioned so as to filter the air entering the underground maze; and at least two species of tuco-tucos are known to maintain burrow temperatures of between 68 and 72° F (20–22° C) by coordinating the opening and closing of tunnel entrances with the wind and sun. Even self-defense is practiced in the burrow, since the eyes of tuco-tucos, situated nearly on top of the head, permit the animals to scan the area above them from the security of the burrow entrance. If danger is imminent, the tuco-tuco simply backs up, guided by its extremely sensitive tail.

North American pocket gophers and tuco-tucos are so closely related that they are considered each other's ecological and structural counterparts in spite of the fact that only the former have cheek pouches. The tuco-tuco's common name is derived from the sound the animal makes while in its subterranean chambers.

Pampas Fox
Dusicyon gymnocercus

Pampas foxes are inveterate collectors and have been known to steal unguarded bits of leather and clothing from travelers. They seem well suited to open country, often preferring to den in the cover of tall grasses or crops rather than seek out the forest. Most dens are "borrowed" from viscachas, armadillos, and other animals, and persistent fox mothers with new litters may force out entire viscacha colonies.

Like some other mammals, pampas foxes "play dead" when they feel threatened. They have even been known to continue to feign death right through a severe beating. No one knows exactly what mechanisms trigger this state, but they seem to reduce the animal's normal sensitivity to pain in some way. In view of the openness of the country and the animal's fairly slow running speed, the adaptation must be effective.

Although classified as a carnivore, the pampas fox has omnivorous eating habits, including a fondness for sugar cane. It preys on small rodents and even on birds and reptiles when it can.

Vicuña
Vicugna vicugna

Weighing only about 100 pounds (45 kg), the vicuña is the smallest member of the Camelidae family, which also includes the llama, the guanaco, the alpaca, and the Old World camels. It is a lovely, dainty-looking animal with small delicate hooves suitable for traversing rocky terrain. This mammal does not inhabit the pampas, but lives on the grassy mountainous plateaus of Bolivia and Peru, to which it retreated in order to escape encroaching civilization. Poor grazing in this dry, cold habitat affects reproduction and growth so severely that in some years up to 50 percent of the young (fetuses as well as newborns) do not survive.

Vicuña wool is perhaps the most luxurious wool of any, but vicuñas are not suitable for domestication, since they will breed only in the wild. This factor has contributed to the animal's exploitation, particularly since poaching, though officially outlawed, is often not suppressed. A female vicuña will stand over the body of a dead male, thus making herself an easy mark for hunters, and entire herds of these pretty mammals have been obliterated in recent years.

6

The Desert

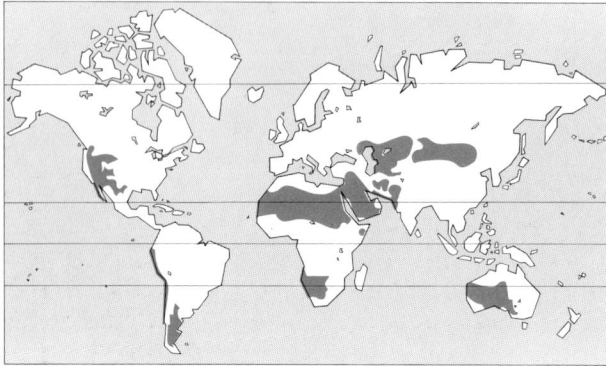

"Pared down" describes the various desert landscapes. Low, widely spaced plants give the terrain a stage-set look. Buttes and mesas lavishly colored through the oxidation of minerals appear stark against the sky. The lines and swirls of hot dunes suggest a primitive kind of finger painting, a giant, slow sculpting and resculpting by the wind.

Such portraits of stillness and simplicity are an important part of the fascination of the desert, but they tell only the daytime half of this biome's story. The other half unfolds in the cool of the evening, when most desert activity begins. Small rodents and rodentlike mammals leave their burrows to forage for seeds, while snakes, owls, and a few carnivorous mammals prepare to attack mice and rats, their chief prey. The browsing herbivores, most of them ungulates, make their way to the twigs and leaves of sparsely scattered plants. Or, if the memory of rain is particularly dim, they may unearth some moisture-rich bulb or other source of water. The paces and rhythms are varied, but every desert animal's purpose is the same—to avail itself of food and moisture before livable temperatures once again vanish in the paralyzing heat of noon.

The enormous difference between the desert's midday highs (120–130° F [49–54° C]) and the cool temperatures of evening (60–70° F [16–21° C]) is due to very low humidity and a correspondingly high rate of evaporation. The desert sky, virtually cloudless, allows up to 90 per-

cent of the sun's rays to penetrate the ground by day. That same cloudless sky allows the day's warmth to dissipate quickly at night—without a layer of humidity there is no insulation of the earth's surface, and bitter cold may result.

In the desert biome, lack of water is of even greater importance than the widely fluctuating daily temperatures. By strictest definition, a desert is an area that receives, on the average, less than 9.75 inches (25 cm) of rainfall a year. That is so little that water, or the lack of it, may be said to determine virtually every aspect of desert life.

Much desert land is found in the interior of continents, or on the lee side of coastal ranges that drain the sea air of its moisture before it can travel far. Some deserts do extend nearly to continental shores, but because deserts are aligned roughly along the tropics of Cancer and Capricorn, their coastal position affords them little rain. (In these latitudes ocean currents are often cold, and they chill the winds that sweep over the land, instead of warming them and producing rain.) When precipitation does occur, it is often in the dramatic and short-lived form of a thunderstorm, after which drought may again prevail for as long as several years.

Remarkably, desert plants are adapted to take advantage of this erratic rainfall. The seeds of many annuals have tough coatings that soften only in rain. Once the coating has softened, the embryonic plant inside can germinate. It quickly matures, flowers (often with acres of identical, brilliantly colored plants), and produces seeds of its own—all before the life-sustaining moisture is gone. Those seeds will then lie dormant in the soil until another rainfall unlocks the life in them. And so the cycle is perpetuated.

Most perennials are adapted through specialization of parts. Some minimize the need for evaporation by losing their leaves during droughts. Others have enlarged roots that hold moisture. And many of the well-known cacti store water in their stems. During dry periods the cacti may draw on this supply of stored moisture until the

Preceding pages: Dramatic panorama
of sandstone mountains carved by the elements
in central Australian desert

entire plant shrivels and the stem's diameter is significantly reduced. Cactus roots are generally shallow, but they extend in a wide radius and so take up much of the little rainwater that is absorbed by the top layers of the soil. It is for this reason, among others, that some desert plants occur so far apart. Some trees do not rely on rainfall at all for their moisture, but send down very long tap roots capable of locating deep ground water. Because they are not competing for surface water, these plants are usually found growing close together.

No one who has ever handled a cactus or similarly prickly desert plant can fail to be impressed with the length and sharpness of its spines. Even these are adaptations to drought. Were the spines not present, thirsty desert animals would no doubt eat the entire plants to obtain their stored moisture, and could wipe out a complete species in a relatively short time. The thorns and briars discourage these animals, however, and an occasional nibble is usually all that can be garnered from the barrel cactus, the prickly pear, and other plants of the desert biome.

Desert mammals are jugglers. When the need to cool their bodies becomes paramount, for example, some of them undergo evaporation, a form of temperature regulation that expends body water. This water loss must be offset by an equivalent intake if the animal is to remain in total water balance, and some mammals meet this need by drinking. But when freestanding water is unavailable, as it often is in this biome, it becomes necessary to find other sources of water. It is these other sources, and the various ways in which the animals take advantage of them, that make desert mammals fascinating, especially since their other adaptations such as locomotion, general size and pallid coloring, feeding habits, and specialized appendages are so closely interwoven with their basic moisture needs.

Among the most typical and intriguing of all desert mammals are the rodents such as the North American kangaroo rats and pygmy kangaroo rats, some of the ger-

bils and jerboas of Asia and northern Africa, the springhare of southern Africa, and the hopping mice of Australia. As their names indicate, these animals are very similar to one another. The Old World desert jerboas and the New World kangaroo rats, in fact, are each other's ecological equivalents, having evolved the same adaptations in response to desert living.

The Merriam's kangaroo rat, *Dipodomys merriami*, has been studied extensively, and serves to illustrate many of these adaptations. Like the larger mammal whose name it bears, this rodent has hind legs that are enormous in proportion to the rest of its body. The advantage these long legs provide, of course, is bipedal jumping power—kangaroo rats are capable of 2-foot-long (61-cm) leaps and can travel some 18 to 20 feet (5.5–6 m) per second. Furthermore, because their long, tufted tails counterbalance their weight, kangaroo rats can actually shift direction in midair. In the desert, where there may be great distances between food and cover plants, and where hungry predators eagerly await their prey, such rapid locomotion and ease of maneuverability are often the keys to survival.

Avoiding predators in the desert's vast open spaces often depends on being warned in time. Kangaroo rats have an acute sense of hearing that helps in this respect. Or at least it is suspected that the very large skull projections that house the middle ear enhance hearing. In addition, the kangaroo rat's tiny size, usually between 1.4 and 5 ounces (40–140 g), permits this rodent to take refuge under rocks and plants where large predators often cannot fit.

To animals that live in hot climates, the benefits of being small extend far beyond the ability to escape predation. Heat loss is maximized in small species, since their body surfaces are very large in relation to their total volume. The advantages of being small enough to burrow (and many desert animals are this small) become obvious when one considers that larger mammals such as the camel have no comparable means of escaping midday heat. They do, of course, have their own very efficient hot-weather adaptations, but these often involve moisture loss. In order to achieve evaporative cooling by perspiring, for example, camels must expend precious water. However, the kangaroo rat (which has no sweat glands) can remain reasonably comfortable during the day simply by sealing itself just a few inches underground, where the temperature is often considerably lower than that aboveground.

Burrows also play a part in the kangaroo rat's astonishing method of conserving water. While sealed in its cool, damp home, this rodent exhales moisture, a small percentage of which is absorbed by the dry seeds that are cached there. This moisture reenters the rodent's body later as part of the food the animal consumes. A greater percentage of the crucial moisture is recirculated more directly, as humidity in the air the rodent breathes. And because it is nocturnal, the kangaroo rat can also take advantage of the relatively high moisture content in the night air outside the burrow.

In a biome where animals may never see rain, drinking as such is quite a luxury, and the kangaroo rat does not indulge in it. What little nonatmospheric moisture it gets on a regular basis comes from seeds, from a few supplementary sources such as leaves and stems, and from the scant amount of metabolic water its body produces through the oxidation of some food elements. Were the animal to excrete liquid urine, almost all of this precious moisture would be lost. Instead, the kangaroo rat's specialized kidneys concentrate the urine into a semisolid state.

Desert carnivores are in a different position, in relation to total water balance, from small herbivorous mammals such as rodents. The body fluids in their prey are a significant source of moisture. Therefore, except when water is necessary to replace that expended in temperature-controlling activities such as panting, desert carnivores should not need to drink. Such thinking is only hypothetical, however, and has not really been tested on wild species. In fact, of all Sahara carnivores, the fennec is the only one known not to drink regularly; jackals and

Top: Aloe trees, South-West Africa. Bottom: Swirling Sahara sands after rainstorm, Niger

other carnivorous species of arid lands are seen very often at watering places.

Red kangaroos and euros, both large marsupial herbivores that inhabit arid lands, expend copious amounts of water through their very particularized form of evaporative cooling. They produce large amounts of saliva, enough to be picked up by the tongue and spread generously over the hot abdomen and extremities. Then they position the wetted-down parts so that they will catch the maximum amount of air. Both of these kangaroos have highly concentrated urine, and the moisture lost in this type of evaporative cooling is not extremely crucial to their total water balance. The red kangaroo has another adaptation that prevents it from overheating quickly. It has reflective fur, which stops solar radiation from penetrating very deeply. The euro, which does not have a reflective coat, escapes the heat by taking refuge in caves.

There is one noticeably well-adapted mammal that is neither a burrower, a carnivore, nor a large ungulate or other type of large herbivore, and that is the jack rabbit. Jack rabbits, like kangaroos, inhabit both the desert and the hot, arid plains, and have a very particularized adaptation that enables them to survive in those biomes. Their extremely large ears, which are laced with blood vessels, radiate heat. In addition, jack rabbits usually rest in depressions in the ground, most often in the shade of some tree or shrub, while the ears are dissipating heat. Were it not for this combination of cooling factors, the animal would probably die.

This is not to say that the jack rabbit is unique in having large appendages. In fact, if one were to line up the arctic fox, the red fox, and the fennec, one could determine, within limits, which biome each of them hails from, simply by looking at the length of the ears: the longer they are, the hotter their owner's native habitat. Whether or not this adaptation operates exactly the same way in all the species that have it is still a matter of some conjecture. But it is safe to say that heat is dissipated more readily from animals with long ears than it is from their short-eared counterparts.

Desert Hedgehog
Hemiechinus sp.

The tympanic bullae—bony structures that house the middle ear—are strikingly large in desert hedgehogs. They probably serve as resonators both for aboveground sounds and for vibrations in the soil where hedgehogs burrow. In addition, two *Hemiechinus* species have conspicuously long ears, which almost certainly act as heat radiators.

H. auritus aegyptiacus has been the subject of recent experiments involving diet and water balance. It is known to be almost totally carnivorous, and seems quite satisfied with the moisture it obtains from insects, toads, small mammals, and other meaty foods. In fact, the water metabolism of this subspecies may rival that of the New World grasshopper mouse, *Onychomys sp.,* an animal known never to require free water. Another desert hedgehog, *Paraechinus micropus micropus,* has even been kept in captivity for up to six weeks with neither food nor drink, and has survived.

When they are frightened or when they are chasing prey, desert hedgehogs can run at more than twice their usual speed. This means that they can travel up to 24.5 inches (62 cm) per second, an astonishing rate for a compactly built insectivore such as this one. The *Hemiechinus* species carry their bodies well off the hot ground when they run, having somewhat longer legs than hedgehogs of the genus *Paraechinus.*

Desert Jerboa
Jaculus jaculus

Like most desert rodents, desert jerboas are incessant burrowers. During burrowing, sand is kept out of their ears by the tufts of hair that surround the ear opening; and it is kept out of the nose by a retractable skin fold over the nostrils. In some jerboas, part of the skull projects in front of the eye cavity and helps protect the eyes as the mammal digs.

J. jaculus caches very few seeds in its burrow, relying instead on fat reserves to see it through inclement weather. (It goes into a torpor during very hot and very cold periods.) After routine nocturnal activities are taken care of, *J. jaculus* plugs its burrow entrance, not with its forepaws as kangaroo rats do, but by using its nose and incisors to tamp the soil. Although sealing the burrow undoubtedly helps to maintain reasonable levels of humidity, it also probably serves to discourage snakes from preying on this tiny rodent.

Spiny Mouse
Acomys cahirinus

Like African ground squirrels of the genus *Xerus,* this spiny mouse species does not estivate or go into a torpor of any sort during summer. It does not even burrow, but inhabits hot, dry rock crevices, where it loses water freely through evaporation. How, then, does the spiny mouse tolerate the dehydration it must experience under such conditions? Recent investigations show that the volume of this rodent's plasma, the fluid component of the blood, remains constant even when dehydration occurs. This is the animal's salvation in extremely arid conditions.

Birth in a colony of spiny mice is a complex social phenomenon. After a gestation of between thirty-five and forty days, the female delivers her two or three well-developed young with the help of other females. Acting as midwives, they may start to clean the offspring even before they are fully born, and in loosening the embryonic pouch, they provide the young with the freedom to breathe and move about. The "midwives" may become very possessive, and in fact often attempt to nurse the young they have helped to deliver. After two or three days of confusion, the colony settles into a pattern of communal care. Young are tended not only by their own mothers, but also by the females that have adopted them.

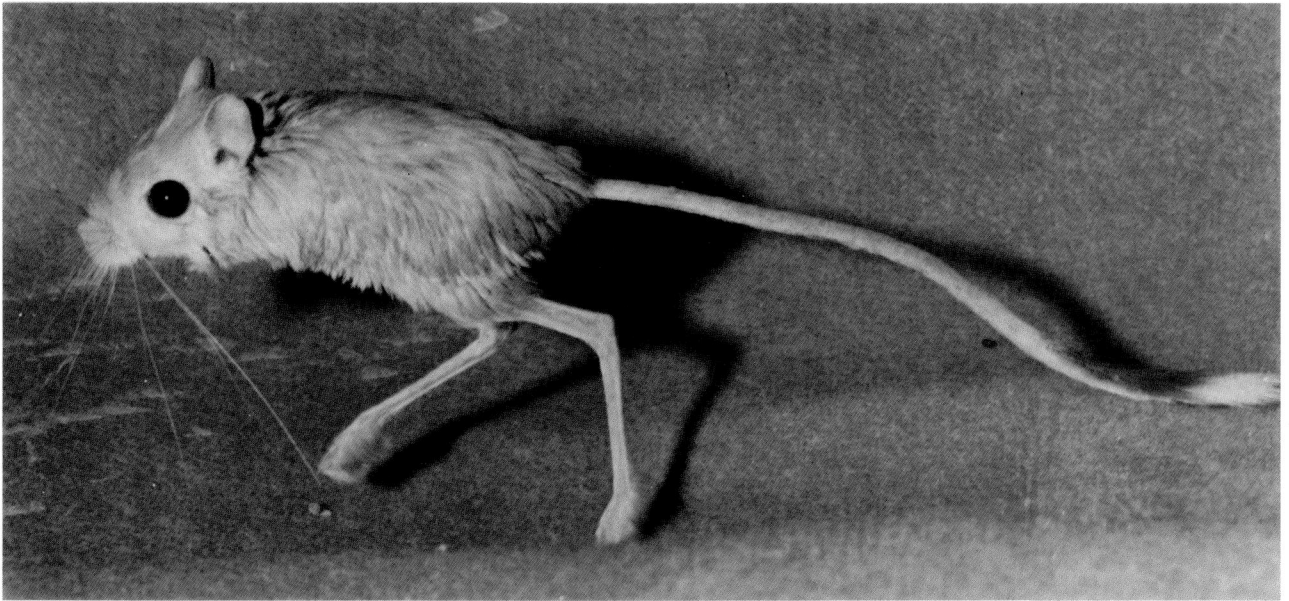

Desert jerboa ▲ *Family of spiny mice* ▼

Springhare
Pedetes capensis

This inhabitant of southern and east-central Africa is a giant among kangaroolike desert rodents, and is not a hare at all. Equipped with tiny, clawed forefeet, springhares are rapid burrowers and have an ear tragus specialized to keep out sand. The entrances to their tunnels are surrounded by loose, sandy soil and usually are plugged from within, except for one opening that is used as an escape hatch. It no doubt surprises a springhare's predators to see this mammal come catapulting from its burrow at twilight in one enormous leap.

Standing and sitting upright present no problem to the springhare, thanks to its enormous hind feet and its tail, which serves as a brace. In fact, springhares sleep while sitting on their haunches with their head tucked between their legs and their furry tail wrapped about them to conserve warmth. They also fold their ears, which are larger than those of any other rodent, when they need to keep in heat. Springhares make 26- to 29½-foot (8–9-m) leaps by using their tail as a counterbalance; they also use the tail to shift direction in midair as they zigzag from predators.

Fennec
Fennecus zerda

Fennecs are small foxlike mammals weighing only about 3 pounds (1.4 kg) each. They burrow so rapidly and deeply in the Sahara and in Arabian deserts that they actually appear to sink into the sand. They can probably survive indefinitely without drinking water, although they are known to frequent oases on occasion. Their very concentrated urine helps minimize water loss, and a somewhat flexible body temperature allows the fennec to become relatively hot before it begins to pant and thereby lose moisture as it cools itself.

The fennec's ecological equivalent in the New World is the kit fox of the Nearctic realm. Like the fennec, it is literally almost all ears. This adaptation, of course, helps to maximize hearing in the vast open stretches of the desert, as well as to dissipate body heat. Though alert, kit foxes are not bold. They are shy, nocturnal carnivores that rely on short bursts of speed to capture rodents, rabbits, small reptiles, and even a few insects.

Sand-dune Cat
Felis margarita

In Arabia and the Sahara the sand-dune cat inhabits areas of constantly shifting sands; in the Kyzyl Kum, the Kara Kum, and other deserts of western Asia, it inhabits dunes with a cover of sand acacia or other desert plants. Maneuvering in such terrain requires traction, which *Felis margarita* gets from the thickly padded soles on its feet. The padding may also insulate the feet from hot sand.

These lovely mammals are small in comparison with other wild felids—usually they measure only about 10 inches (25 cm) at the shoulder. Nocturnal for the most part, they spend the hottest hours of the day in a shallow burrow or under scant vegetation. Their large and widely spaced ears, set low on the head, flatten easily and allow them to crawl under very low-growing plants.

Sand-dune cats, like so many other desert mammals, apparently get all the liquid they need from their food (rodents, birds, reptiles, and locusts). They are relatively rare in zoological collections. However, it is suspected that they occur fairly frequently in the interior of Arabia and in the area south of the Aral Sea in Asia.

▲ *Fennec* ▼ *Sand-dune cat*

Dromedary
Camelus dromedarius

The secret of the legendary camel is in its hump, which is made up of stored fat from the food obtained in plentiful times. During periods of drought the camel draws on this hump for energy. As fat is broken down in the hump, hydrogen is released; this hydrogen combines with oxygen taken in during breathing to produce metabolic water. During drought the hump may shrink and lean to one side. When food and water are available again, the hump swells rather quickly. (Although camels are able to drink an enormous amount of water at one time, they usually drink only when they need to regain lost weight.)

The rise in a camel's body temperature during periods of extremely hot weather allows for heat storage in the daytime. At night, when temperatures are cooler, the camel can rid itself of this stored heat without losing body water in the process. (It perspires only when its internal temperature is 104° F [40° C] or higher.) Concentrated urine also helps in the conservation of moisture. The camel's nostril grooves channel moisture from the nasal passages into the mouth, and the nostrils themselves close to keep out sand.

Only the two-humped Bactrian camel is still found in the wild; however, it is not considered the true desert animal that the one-humped dromedary is.

Addax
Addax nasomaculatus

The addax, an antelope of the Sahara and a close relative of the oryxes, differs from the oryxes primarily in its dramatically curved horns and broad hooves, which are ideal for maneuvering in sand. Little is known of the addax's water metabolism other than that the animal seldom if ever drinks. Apparently it obtains its moisture from sparsely scattered green plants, which it travels considerable distances to reach. In this respect it and other large herbivores have a distinct advantage over small mammals, which cannot travel so far.

Like the camel, the addax is a ruminant, and the large amounts of digestive fluids both animals produce have been ingeniously harvested and used by human beings, either to provide liquid for themselves to drink, or to give to their other animals. Natives of the Sahara lift the rumen from the body of a slaughtered addax and set it on sticks that have been placed between the animal's horns. (The head, obviously, is first severed from the addax's body.) Then they position the animal's skin beneath the dripping rumen to catch the liquid.

Unfortunately, addaxes are slow runners, and many of them have been tracked and killed by hunters.

Aoudad, or Barbary Sheep
Ammotragus lervia

Africa's only native wild sheep, the aoudad is a ferocious fighter. During rut, males often lock their curved horns and try to pull each other down, or they rush toward each other and clash horns violently. But their marvelous strength and pride have failed to save them from natives who hunt them for their meat and hide.

Too large to take cover behind most desert plants, aoudads will freeze when threatened, and thus blend with their barren habitat, their coloration being perfectly suited to this ploy. Excellent jumpers and sure-footed climbers, they roam the mountains during the summer in search of green vegetation. In winter, dry grasses and lichens supply moisture.

Like most open-country species, aoudads give birth to well-developed young, which learn very early to climb and jump as skillfully as their parents. Although they continue to nurse for about six months, they begin eating grass when only one week old.

* Dorcas Gazelle
Gazella dorcas

For many gazelles, reproduction seems to be related to rain: the calving season begins about thirty days after the start of spring rains, when green food plants are plentiful. Gazelles also seem to be adapted to desert living through some heat-rejecting property of their coats. However, much remains to be discovered about this, as it does about the gazelle's ability to survive in the desert without any apparent means of storing fat.

Only about ten to twelve species of the genus *Gazella* are recognized, although the word "gazelle" is sometimes used to describe other graceful antelopelike ungulates as well. Not all of the true gazelles are desert-dwellers. The fawn-colored dorcas gazelles are small, only about 2 feet (61 cm) tall at the shoulder. They inhabit the dunes and flat rocky areas of most of northern Africa. Acacias, locusts, and succulents are their standard foods, and they travel far on their long, slim legs to reach them.

Three subspecies of dorcas gazelles are endangered, due to serious overhunting and to habitat degradation caused by overgrazing by domestic livestock.

Marsupial Mole, or Pouched Mole
Notoryctes typhlops

Marsupial ''moles'' are not related to moles of the insecti-vore order. They resemble those mammals, the golden moles in particular, because over the millennia the pres-sures of a common way of life caused their physical de-velopment to proceed along parallel lines—another exam-ple of convergent evolution. The fossorial, or burrowing, life style of these marsupials, however, differs from that of the true moles in one rather peculiar way: they virtually swim through the soil without leaving a permanent bur-row behind them. After tunneling several feet they emerge and shuffle along aboveground for a short dis-tance. Then nose, claws, and feet rapidly propel them back into the sand.

Only the duck-billed platypus was as exciting a dis-covery to early mammalogists as was the marsupial mole. Little is known of the breeding habits of these marsupials, as they are difficult to observe in the wild.

Euro, Wallaroo, or Hill Kangaroo
Macropus robustus

Kangaroos generally occupy the same niche in the world of marsupials that grazing ungulates such as deer and antelopes do in the world of placentals—they are large, fleet-footed herbivores. Euros require nitrogen in order for their stomach bacteria to break down the cellulose in certain food plants. But most plants of the Australian des-ert are low in nitrogen. Marvelously, the euro is adapted to obtain nitrogen from its own urea, and it recycles this substance through its digestive tract rather than excreting it in liquid urine. The process, of course, also conserves considerable moisture.

Euros are considered pests in domestic sheep pas-tures, and some livestock farmers have tried to eliminate them by poisoning their water supplies. Interestingly, however, they cannot poison an entire population, since it is quite possible that some euro individuals drink and others do not. In calculating poison dosages it was deter-mined that those that do drink consume only 1.7 percent of their body weight per day in liquid when outside tem-peratures are relatively low. But that figure rises to 5 per-cent when the animals are restricted from eating fresh foods or cannot escape high temperatures by retreating into caves and rocky crevices.

Rabbit-eared Bandicoot, or Bilby
Thylacomys sp.

On hot summer days rabbit-eared bandicoots escape the heat by retreating to their unusual spiral-shaped burrows, which may be some 6 feet (2 m) deep. To excavate them, bandicoots first loosen the soil with their well-clawed forefeet, then throw it backward with their strong hind legs. Communal living is not their style: each burrow is occupied by a single individual or at most by a pair or a female with young. However, burrows are fairly close to-gether. In certain areas bilbies occupy them throughout the winter. They sleep in a squatting position with their head tucked forward and their long ears folded over the eyes and part of the face.

This silky marsupial is so skilled at catching insects and mice that at one time it was kept as a pet for that pur-pose. Recently it has suffered the effects of poisoning, hunting (for pelts), and habitat changes. Now it is consid-ered rare.

Two species exist. The common rabbit-eared bandi-coot, *T. lagotis,* is restricted to inland deserts and parts of southern Western Australia. *T. leucura,* the smaller species, lives in the central part of the continent.

▲ Rock wallaby ▼ Dingo

Rock Wallaby
Petrogale penicillata

Australia's answer to the mountain goat, rock wallabies make startling leaps between boulders or cliffs up to 13 feet (4 m) apart. And they remind onlookers of the tree kangaroos in their ability to scale leaning trees. Their well-padded and somewhat rough hind feet provide good gripping power, and the friction they create wears regularly used rock surfaces glossy and smooth. Unlike many similarly built marsupials, rock wallabies do not use their tails as props. Instead, the tails provide rudder action and serve to balance the animals during leaps.

The rock wallaby is widely distributed throughout Australia, wherever caves and fissures are associated with rocky terrain. The constant temperatures of these retreats probably protect the animals from outside temperatures that may fluctuate as much as 50 or 60° F (10−16° C) in a day. The rock wallaby's retreat to the caves is comparable to the burrowing of smaller animals, and the adaptation is shared by the euro as well. Although these small marsupials (they weigh between 7 and 20 pounds [3−9 kg]) are nocturnal, they will leave their daytime haunts and sunbathe if temperatures are not too hot.

Dingo
Canis dingo

The dingo is Australia's only native terrestrial carnivore that is not a marsupial. What a contrast this is to the abundance of nonmarsupial carnivores found throughout the rest of the world, where oceans did not restrict the free interchange of mammals in ages past!

Although they resemble other members of the dog family, dingoes have some unique features: their constantly erect ears, their yelping (rather than barking) sounds, and their brushlike tail. They also have larger canine teeth than domestic dogs, and seem both more affectionate and, at the same time, more withdrawn than domestic species. Dogs can interbreed with dingoes, but it is doubtful that they do so in the wild with any regularity.

Though the fossil record shows that dingoes were most probably brought to Australia many centuries ago, they are usually considered natives. Dingoes are a serious threat to domestic sheep that graze in parts of their nearly continent-wide range, and they are often caught in traps set with poisoned bait.

Australian Hopping Mouse
Notomys alexis

Australian hopping mice are prototypical desert rodents, being adapted for rapid bipedal locomotion in the style of the kangaroo. They are known to be able to subsist with no free water, to have highly concentrated urine, and to go underground during the hottest times of day. Perhaps the most interesting thing about them is that until recently, they were thought to have a marsupial counterpart with which they shared quarters. This kind of cooperation is practically unheard of among species that are actually each other's equivalents, or very nearly so. High-speed photography has revealed, however, that the so-called marsupial counterparts (mice of the genus *Antechinomys*) are not counterparts at all, since they are adapted to move on all fours instead of on the two hind limbs only.

Most hopping mice have an unusual type of tissue on the throat or the chest. Also noted in certain bats, it contains specialized glandular arrangements and hairs, but no one seems to know the patch's function.

Mongolian Gerbil
Meriones unguiculatus

Ever since 1935, when twenty pairs of Mongolian gerbils were captured, the demand for these rodents as pets and laboratory animals has been high. Much of their popularity as pets rests on the fact that, since they are species of both desert and arid steppe, the sociable gerbils are quite satisfied with a diet of dry food. In captivity they seldom bite, and in fact have no habits more troublesome than that of rhythmically drumming their hind feet on the ground as some type of signal to their fellows. (Kangaroo rats, another species of open spaces, also communicate by drumming, with one hind foot at a time. Deer mice drum with the front feet only.)

Meriones is only one of several genera of rodents known as jirds, gerbils, or sand rats. Other genera include *Gerbillus,* with fifty-four species in Africa, Iraq, and Pakistan; *Psammomys,* with two species known as fat sand rats; and the African karroo rat, mole-rat, and naked mole-rat genera. All of these rodents differ from jerboas in not being primarily bipedal. However, they can stand on their hind legs to get a better view of the possible dangers around them. This adaptation is a tremendous advantage to mammals of open spaces, where hiding places are few because of scant cover.

Asiatic Jackal
Canis aureus

Jackals are very wolflike mammals, being omnivorous hunters and scavengers. In fact, they have been almost eradicated from part of Bulgaria by poison intended for wolves. They hunt in packs by night, often running as fast as 35 miles per hour (56 kph). Or, given the chance, they may take the leftovers of a leopard's kill. On very hot days jackals have been known to spend hours resting in pools of water, or simply to search out a thicket.

Like the opossum and several other animals, jackals defend themselves by feigning death. They actually seem to go into a less-than-conscious state, with their heartbeat slowed. No one seems to understand the exact mechanism that triggers this reaction.

All three species of jackals may be found in northeastern Africa. The Asiatic jackal ranges from central Africa to Russian Turkestan; the other two species, *C. mesomelas* and *C. adustus,* inhabit Africa only. In certain Arab countries where the human dead are interred only slightly beneath ground level, heavy stones must be placed on the graves to prevent jackals from digging the bodies out.

Mongoose
Herpestes sp.

Mongooses range throughout many habitats in the Old World, from arid, sandy country to marshes and open forests. In all of them the chief survival tool seems to be their remarkably fearless slaughter of a wide variety of prey animals, most spectacularly snakes and rats. No longer thought to be totally immune to snake venom, mongooses are now known to avoid snakebites through agility and cunning. In fact, staged battles between mongooses and snakes are common tourist attractions in India. Less well known is the alleged fact that mongooses perform antics such as handstands and curious dances in order to attract poultry and perhaps other animals. As one might expect, such eager carnivores can easily become pests, especially after they have accomplished what so many of them are exported to islands to do—that is, to catch rats. Then their attention too often turns to domestic species.

Mongooses treat the eggs that they eventually eat almost like toys. With their front or hind legs, they hurl them repeatedly against rock walls until the eggs are shattered. Often this innate behavior becomes incorporated into play sessions such as sexual play and mock fighting.

▼ Mongoose   ▲ Asiatic jackal

* Kulan, or Mongolian Wild Ass
Equus hemionus hemionus

Large numbers of kulans once inhabited the Gobi Desert and other arid areas of central Asia. The herds have been reduced to practically nothing through unregulated and persistent hunting, most of it done on horseback and from automobiles. The Asiatic wild ass, *E. hemionus,* is now classified as vulnerable, and two subspecies, *E. hemionus khur* and *E. hemionus hemippus,* are endangered.

Ironically, kulans are capable of outrunning domestic horses, wolves, and most other animals. Their average running speed is about 30 miles per hour (48 kph), and their top speed is greater than that of the swiftest Thoroughbred. Extraordinarily long bones in their lower legs account for this amazing speed and are the chief feature that distinguishes the members of the *Hemionus* subgenus from all other forms of wild horses, mules, and asses.

Paler in color and slightly smaller, kulans otherwise resemble the Tibetan kiang, also a member of the *Hemionus* subgenus. The kiang ranges through the Himalayan valleys. Possibly because it breathes thinner air, it has a larger chest than the kulan.

* Goitered Gazelle
Gazella subgutturosa

Goitered gazelles escape the cold of their Asian desert habitat by migrating to find food. Herds may travel great distances in winter, across the steppes or down from the mountains; in summer they return and eat the new vegetation even before the snow that covers it has melted.

Mating and reproduction are also determined by the cold. Unlike the more tropical gazelle species, goitered gazelles have a single fixed mating season. Bucks join herds of does at the beginning of autumn rut, and each of them takes possession of three to five of the females. Mating is usually completed by the end of December. The two or three young that each female bears arrive in early spring, and so do not have to contend with devastatingly cold weather until they are past infancy.

This species' common name is derived from the enlarged throat and goiterlike neck that form in the males during breeding season. This same characteristic distinguishes males of the genus *Procapra,* the so-called Mongolian gazelles.

The sand gazelle, *G. subgutturosa marica,* is an African subspecies that has become endangered through hunting and habitat degradation.

Free-tailed Bat
Tadarida brasiliensis

Tadarida bats are famous for their twilight exodus from caves in the United States' Southwest. In summer millions of them depart from their roosts in droves; they sound like a roaring waterfall as they move, and from a distance they look like a column of smoke rising over the land. They are out to feed—primarily on moths, which they catch on the wing. At Carlsbad Caverns National Park they are thought to fly no more than 40 or 50 miles (64–80.5 km) from home before they return, at dawn, to their caves.

In winter these bats either hibernate or migrate. Those individuals that hibernate withstand desiccation extremely well. The vast majority migrate, many of them to Mexico, making round trips of up to 2,000 miles (3,220 km) to reach their destinations.

The free-tailed bat's common name is derived from the fact that the end of its tail is free of the interfemoral membrane rather than enclosed in it.

Black-tailed Jack Rabbit
Lepus californicus

Long ears, a trademark of the lagomorphs, are of exaggerated length in the jack rabbit. In addition to aiding hearing, long ears help the jack rabbit control its temperature by radiating heat to the environment. They give the animal its common name, which is shortened from jackass.

These lagomorphs are true hares (rabbits differ from hares in structure and behavior). In both grasslands and deserts they are active much of the day. Because of the pale coloration characteristic of these and other species of arid regions, jack rabbits blend readily into their surroundings and are often overlooked by coyotes and other predators. If a chase does erupt, the jack rabbit will bob across the landscape in leaps of 14 feet (4 m) or more; yet it may become an easy mark for patient carnivores since it travels a circular course.

The rabbits and hares of Europe and both the black-tailed and the white-tailed jack rabbit of North America share the habit of reingesting their soft fecal pellets (they produce hard ones also). This probably assures them of the proper vitamin intake, and it reduces food-gathering time. Like the Old World rabbit in Australia, the black-tailed jack rabbit periodically experiences dramatic population peaks, and its enormous numbers during such times make it quite a pest.

Desert Wood Rat
Neotoma lepida

Shiny tin cans, small leftover camping gear, cacti, rocks, and a wide variety of other objects are favorite possessions of desert wood rats, and their nests are often bulging with these "ornaments." The collecting habits of this rodent have earned it the nickname "pack rat."

Pack rats are among the few desert mammals that seem unharmed by *Opuntia* cacti, which include the spiny prickly pear. This assures their nearly exclusive access to those plants, an important source of moisture which they eat in times of drought. Pack rats also pile prickly plants at the openings of their burrows, thus creating predator-proof retreats.

N. lepida is one of the desert species on which studies have been made concerning coloration. Local populations of wood rats may be dark, light, or intermediate in coloring, depending on the type of soil or rock they inhabit. These substrate races, as the populations are called, are thought to have resulted from predators capturing the least well-camouflaged individuals, and thereby eliminating them from the gene pool. Thus, protective coloration operates on a very localized basis in the desert.

Antelope Ground Squirrel
Ammospermophilus leucurus

Antelope ground squirrels are unusual small desert mammals—they neither hibernate nor estivate, but face desert conditions year-round. Their main key to survival under such stress is their use of the burrow. On very hot days the body temperature of these animals rises so high that they are forced to go underground. There they lie on their stomachs long enough to give off body heat to their cooler environment. Then they return to their diurnal activities—searching for seeds, fruits, insects, and even carrion. Because they pop in and out of the burrow, rather than using it over long periods of time at once, antelope ground squirrels actually cope with a fairly low mean ambient temperature, one that is somewhere between the outdoor temperature and that inside the burrow. In addition, antelope ground squirrels often avail themselves of the branches of nearby shrubs, where the air circulates relatively freely. And they are thought to have a thick epidermis that may help to check water loss and to shield solar radiation from their bodies.

Pocket Mouse
Perognathus penicillatus

The tiny pocket mouse has evolved a particularly efficient means of coping with desert heat—it simply "drops out." Estivation, a period of hot-weather dormancy, is practiced by this and a few other mammals (most of them small) as one way of surviving extreme desert conditions. As in hibernation, the animal's body temperature drops drastically; yet the torpor itself does not appear nearly so deep as that of a hibernating mammal, probably because the burrow temperature is still relatively high.

Furred cheek pouches, which give this rodent its common name, open directly onto the exterior of the animal's cheeks, not into its mouth. These pockets are hurriedly stuffed with seeds, which are then emptied into the burrow and eaten at the rodent's leisure. That way, predators are not invited to make a meal of the pocket mouse as it feeds.

Unlike kangaroo rats, pocket mice do not have huge stiffly haired hind feet adapted for hopping. All four feet are of more or less uniform size, and locomotion is more of the scurrying type usually associated with mice.

Javelina, or Collared Peccary
Tayassu tajacu

The name javelina, from the Spanish word for "javelin," refers to this mammal's long, spearlike tusks. These specialized teeth are desert tools *par excellence*—they serve the javelina as weapons against jaguars and other enemies, and they unearth moisture-laden roots and bulbs from hard-baked soil after the javelina has sniffed them out. Water is essential to the peccary in spite of this diet, and bands of five to fifteen individuals often stop at watering holes and streams as they travel. If an insurmountable threat arises, the peccary sounds an alarm, discharges its strong-smelling musk, and wheels off with its fellows. The musk, ejected from a gland located on the animal's back a few inches in front of the tail, is also used to mark territories.

This New World mammal differs from the true pigs, which it resembles, in the structure of the bones in its feet and in the complexity of its stomach. Its closest relative is the white-lipped peccary, *T. pecari*, found in the tropical forests of Central and South America. The white-lipped peccary is both larger and more gregarious than its desert counterpart.

▲ Javelina ▼ Pocket mouse

Mara, Patagonian Cavy, or Patagonian "Hare"
Dolichotis sp.

Maras are strikingly long-limbed rodents. They travel by hopping, by walking, and most often by stotting—bouncing stiffly on all four legs. They can move as fast as 19 miles per hour (30 kph) in this manner. This gait, used by other mammals as well, may have a warning signal built into it, since it causes the mara's white rump to jostle conspicuously. Thirty or forty maras traveling together this way present a striking flash of white across the desert.

When they rest on their haunches maras keep their forelegs straight, and when they lie down they double the forelegs under the chest, catlike. These postures are unusual for a rodent, as is the sitting posture that both mother and offspring adopt during nursing sessions. Perhaps these postures, like the stotting gait, are some form of desert adaptation not yet understood by human beings.

Owners of domestic guinea pigs may be surprised at how little their pets resemble the closely related maras. More typical wild guinea pigs, or cavies, are the moco, *Kerodon rupestris,* and two genera of cuis. The little-known African gundis, which resemble guinea pigs, belong to a separate family.

Peludo, Pichi, and Pink Fairy Armadillo
Chaetophractus sp., Zaedyus pichiy, Chlamyphorus truncatus

Peludos, pichis, and pink fairy armadillos—all are hairy inhabitants of arid South America. The last is the smallest armadillo of all: it measures only about 5 inches (13 cm), exclusive of the tail. Its pinkish armor is in two tenuously connected sections, the smaller, rear one almost perpendicular to the armor that runs the length of the body. This rear plate serves as a burrow plug for the fast-digging pink fairy, and so protects the animal when it is in a vulnerable, head-down position.

Peludos, or hairy armadillos, also burrow as a means of defense, and one species even secures itself in its home by wedging the edges of its armor into the burrow walls. Peludos have been known to kill snakes by cutting them with the edges of their shell. When cornered, they may simply pull their feet under them and let the edges of their armor touch the ground.

Pichis, armadillos that employ some of these same defense methods, are very common throughout their range. Some individuals are made into pets, and many are prized for their tasty meat.

Guanaco
Lama guanacoe

Guanacos, vicuñas, and their relatives the domesticated llamas and alpacas are all members of the Camelidae family. Like camels, guanacos can withstand extreme dehydration; and they recover rapidly, since they can drink enough water in ten minutes to replace one-fifth of their total body weight.

Guanacos exhibit a strong herd instinct, and since they inhabit open country speed is essential to their safety (they can run up to 34 mph [55 kph]). If they find a cool mountain stream in their wanderings, they make a pastime of lolling in it. The guanaco's range is dramatically varied ecologically. It includes the Atacama Desert, the driest in the world, as well as the wet Fuegian Archipelago. And in altitude it extends from high Andean plateaus all the way to sea level.

The importance of the domesticated forms to the economy of South America can hardly be overstated. The quality of alpaca wool is unequaled, and the llama provides products as varied as leather, fuel (from excrement), and rope from the hairs.

7

The Tundra

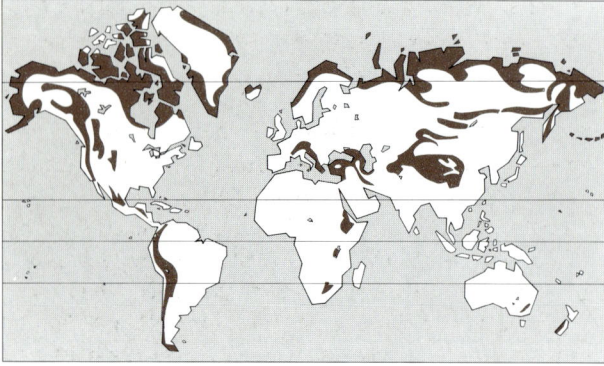

The tundra rings the North Pole almost continuously between the icy Arctic Ocean and the tree line (the northernmost limit of tree growth). From afar, particularly, the look of this land is like the name so often used to describe it—barren ground. The word "tundra," in fact, is derived from a Lapp word that means "rolling plain without trees." Trees cannot grow here because of the low temperatures and the resultant brief growing season. In winter, thermometer readings of −58° F (−50° C) are not unusual, and summer temperatures, which may reach 59 or 60° F (15−16° C) are not sustained long enough to allow tree cells to establish and maintain efficient growth patterns. (Tree cells operate well only in temperatures of 50° F [10° C] or warmer.) Even if temperatures were higher, the bits of rock and sand that tundra winds transport would bombard the tender parts of upright trees. Therefore those few trees that do get a roothold eventually come to hug the ground in self-defense.

Precipitation, evaporation rates, and light intensity are all characteristically low in the arctic tundra. Annual precipitation averages only 6 to 8 inches (15−20 cm), most of that in summer or early autumn rains. These are desert conditions, particularly since the cold air holds so little moisture. At certain places in the Arctic the extreme angle of the winter sun erases true daylight for a few months. And in summer, when the sun may be well above the horizon for many days consecutively, the light

intensity is low compared with that of the tropics. Even so, green plants in the tundra do not lack light for photosynthesis, since the most efficient of them utilize only about one-twentieth of the available sunlight. The main problem for plants, and the reason their growth is both so specialized and so labored, is that the growing season is extremely short, usually only two to three months.

Permafrost is the name of the subsurface layer of ground that remains frozen year-round in the arctic tundra. In parts of Alaska it is some 2,000 feet (610 m) thick; in parts of Siberia, nearly a mile (1,610 m). Above it rests a layer of soil called the active layer. This is the soil that thaws in summer, and in many places it is only inches deep. The interplay between the permafrost below and the sensitive active layer above may account for the tundra's changing face, since permafrost prevents water from being absorbed into the soil, with many consequences. For example, thaws that quicken the active layer may also cause it to shift, and entire slopes may slide gradually toward the valleys from the weight of the water with which they are saturated. The lakes and bogs that dot the tundra landscape so plentifully in summer are actually rainwater or melted snow that the permafrost holds on or near the surface of the earth. (Permafrost does keep water where it is available to some plant roots.) Even the geometric patterns in soil and the cone-shaped mounds known as pingos are the results of permafrost pressure causing the active layer to buckle.

Arctic soils in general are products of repeated ice action. At least three times during the last million years (very recently, in geological terms) great sheets of ice have bulldozed the land and denuded much of the northern hemisphere of its soil layer. Whereas soil in the warmer areas became stabilized after the glaciers retreated, tundra soils were prevented from maturing by repeated freezes and thaws and by terrific wind abrasion. Today much of the tundra is still covered with till, the glacial leftovers that frost action keeps churning about, winter after winter. Because drainage is poor, this soil is extremely acidic, thus hardly conducive to growth.

Preceding pages: Group of ibexes on an Alpine slope

Arctic soils are also characterized by their low nitrogen content. Nitrogen is generally plentiful where bacterial action occurs freely, and cold temperatures in the Arctic keep bacterial action (decomposition) to a minimum. Furthermore, decomposition is slow or nonexistent simply because often there is nothing to decompose, huge stretches of tundra being void of significant animal and vegetable life. Whether low nitrogen critically limits plant growth itself is still not certain.

Given the poor quality of much tundra soil, it is surprising to find vegetation here at all, and indeed the variety of plant life is limited. What thrives in the tundra are communities of low-growing willows, lichens, mosses, grasses, flowering herbs, and, in the better-drained regions, stunted trees. Most of these plants are perennials that reproduce via runners or some other vegetative part. The colors of many tundra plants, such as the scarlet bearberry, are brilliant. Even the mosses and willows may be richly hued, and large patches of lavender lupine, or even white cotton grass, are an esthetic bonus along stretches of otherwise brownish terrain.

Interestingly, some arctic plants not only trap warm air in their dense cushion- or rosettelike growth formations, but also control their own temperature through pigmentation. It has been shown, for example, that the temperature inside blue flowers can be about 4° F (2.2° C) higher than that inside white ones growing in identical conditions. And heavily pigmented plants may even absorb enough light to begin growing under the snow in spring. This may give plants as much as a two-week head start on growth, which is substantial when one considers that the entire growing season in the Arctic may be as brief as eight weeks.

Furthermore, the amount of pigmentation in a given species may vary dramatically among individual plants or colonies of plants. In many species this variation seems to be environmentally controlled, since a plant nestled in the warmth of some protected site will be lighter in color than another individual of that same species growing in a totally exposed area. What a finely wrought adaptation

to this severe biome, and how well it illustrates the seemingly minimal adjustments that can make life-or-death differences to tundra species.

After the bulldozing Pleistocene glaciers receded for the last time in the northern hemisphere, there appears to have been a continuous cover of tundra vegetation connecting Old and New World lands, and extending well into the Arctic. As the climate warmed, forests replaced tundra vegetation in the southern lowlands. But on the higher mountains, in the cold conditions above the tree line, tundra vegetation still exists today. This mountaintop biome is called the alpine tundra, and alpine tundras are found between 72° N and 56° S latitudes.

Although some alpine and arctic plant species are identical, differences between the two types of tundra are important to an understanding of the mammals that inhabit them. Most obviously, alpine tundra is discontinuous, being dotted over the globe except where it meets arctic tundra in the North. This discontinuity naturally discourages animal interchange. Thus, alpine mammals in various parts of the world are very different from each other. In the arctic tundra, on the other hand, many mammals are present both in the Old and the New World, and are therefore described as circumpolar mammals. (Because arctic-tundra mammals are so distributed, they are not grouped according to continent here.) Simple geographical access and similarity of habitat all around the Arctic Ocean are no doubt responsible.

As mountain climbers know, the atmosphere of the alpine tundra is considerably thinner than that of lower elevations. Extremely long periods of daylight and dark are not a factor in alpine areas, nor is permafrost. The extreme inclines, in fact, encourage runoff, an important characteristic since precipitation in the alpine tundra is relatively high. And the fact that scant vegetation occurs on slopes rather than on level ground has produced an array of alpine-tundra mammals equipped to forage, and in some cases even to bear their young, on the rocky inclines of mountains, near the top of the world.

Alpine- and arctic-tundra mammals share many traits, regardless of habitat distinctions. For one thing, few of either group seem to have gravitated directly to the tundra, over the millennia, without having resided in other biomes first. It is anyone's guess whether the alpine mammals' steppe heritage exerted some sort of pull to the tundra lowlands. Or perhaps competition in the temperate zones simply forced many species to take to the hills, or to the North. Even now, the habitat of some mammals such as the caribou is determined by the season, since they feed in the polar realms when food is plentiful, then return south in autumn. Others adhere to a life style that allows them to inhabit either tundra or taiga almost year-round. Wolves and wolverines, for example, are so adapted, and consequently are somewhat difficult to place ecologically.

The number of tundra mammals is small, since the scarcity of lichens, mosses, and other food plants in this harsh environment limits the number of herbivores. Carnivores, of course, feed on herbivores; therefore, as second-order consumers, they too are affected by the sparse vegetation. The phenomenon of cyclical fluctuations in mammal populations, seen most spectacularly in tundra lemmings, is related to this scarcity. Every three to four years in some species, every nine to ten years in others, the herbaceous food supplies are simply not great enough to feed the bumper crops of small herbivores. As a result, many starve. The dying out of these prey species eventually causes a famine for large numbers of predators—the foxes, wolves, and other animals that depend on the rodents for food. Not until the vegetation, the prey species (herbivores), and the predators somehow strike a balance once again do the affected populations become restabilized.

Scientists are hard pressed to explain these mysterious fluctuations. Perhaps they are actually a long-range survival adaptation for the very species that are destroyed in their wake, since only the best-adapted individuals will survive to reproduce. It has also been suggested that pre-cious nutrients such as phosphorus and nitrogen, usually locked up in plants if present at all, are made more widely accessible to the tundra through the droppings of the multitudinous herbivores, or through their entire bodies: if the herbivores do not provide carnivores with food directly, then their decomposition may nourish the soil for future plants and hence for future herbivores (prey). Whatever the explanation, it is almost certain that mammal populations fluctuate dramatically in the far North (the tundra and taiga) not by any quirk of nature, but because somehow the phenomenon contributes integrally to a carefully balanced ecosystem.

Specialization is obvious in many tundra mammals, particularly in the structure of their feet. Arctic species such as polar bears and caribou possess huge feet that enable them to move easily on snow, ice, or flat boggy terrain. The larger alpine herbivores such as certain members of the goat and sheep clan inhabit country where raw rock outcroppings and jagged inclines abound. They are, necessarily, among the most sure-footed of mammals, thanks to their specialized hooves—sharp on the edges and concave in the center of the underside. This arrange-

Saxifrage, wild herb
typical of the arctic tundra,
Victoria Island, Canada

ment creates a suction-cup effect under pressure and provides exceptional agility.

Cold is, of course, a condition that all tundra species must reckon with. Many, such as the caribou and wild sheep, are adapted to cold through a relatively large body size. Heat loss is minimized in large animals because their total body surface is small in relation to their weight (volume). Conversely, mammals living in extremely hot climates, where heat loss is desirable, are generally small.

There are some small tundra mammals, of course, and they often minimize heat loss through small appendages such as tails and ears. (Extremities tend to radiate heat. Therefore the smaller they are, the less heat they radiate.) Even tucking their extremities into or under the bulk of the body helps mammals keep their heat loss to a minimum. And although snow (a good insulator) is swept away by the strong winds, small drifts that accumulate around the bases of plants often provide pockets of relative warmth in which small mammals can find shelter.

Some tundra mammals have fur on the soles of their feet, and the winter pelage of foxes and wolves is much thicker than their summer pelage. The character as well as the density of many larger polar mammals' coats helps to cut down on heat loss. For example, coats are usually double-layered, with an outside layer of oily hairs that provide effective waterproofing and an inner layer of hairs that trap heat in air pockets. Or, in some species without the undercoat, outer hairs themselves are much thicker near the ends than at the bases, and they overlap to hold in body heat. Layers of fat, of course, function as insulation in some arctic species in much the same way that fur does in others.

The white coloration of many arctic-tundra mammals may be some sort of heat regulator. The winter sun's infrared rays are absorbed equally well by dark and light coloration. But white mammals seem to lose less heat through radiation than dark ones do, though the subject is still a controversial one. The white coloration of tundra species is often considered a response to several differ-

ent stimuli, the need for camouflage in snow being one of the most important.

Cold-weather mammals have some control over the amount of heat that their bodies actually produce. Metabolic rate does not increase in Dall's sheep, for instance, until the temperature drops below $-22°$ F ($-30°$ C). Above that point Dall's sheep in their winter coats are, if not comfortable, at least not dangerously cold. Other species, reindeer for instance, have considerably lower "normal" temperatures in their limbs than in the rest of their bodies. The tissues in these exposed parts are capable of functioning with less heat than the other tissues require.

Related to heat production is the problem that some polar mammals have with movement. Although the heat generated by rapid movement would seem to be desirable, polar mammals cannot always afford to move fast, because perspiration is a liability. Evaporating perspiration can chill the body dangerously, and the frozen moisture may even reduce the insulating properties of fur.

In spite of marvelously efficient body coverings and metabolic adaptations, many tundra mammals are simply not equipped to withstand the combination of severe cold and impoverished winter food supplies. Such mammals migrate or hibernate (or go into a torpor similar to hibernation but not nearly so extreme). Those mammals too small to travel great distances (voles, for example) take refuge in the relatively warm and constant temperatures beneath the snow.

To animals that manage to survive the long tundra winter, the arrival of spring is a rich reward. In arctic regions it is accompanied by much hustle and bustle, since a great deal of living must be packed into the two or three frost-free months. Once solar radiation is absorbed into the ground in spring, the soil temperature begins to rise, and it usually reaches figures above freezing as much as a month before the air does. In many cases it is critical that tundra mammals take advantage of this change in soil temperature in whatever ways they can, especially since their May offspring must be past infancy before the winter overtakes them.

Arctic Shrew, Siberian Shrew, and Masked Shrew
Sorex sp.

Shrews are frantic insectivores that do not hibernate, but hunt and eat nearly constantly. Their tiny bodies (they often weigh less than half an ounce [14 g]) have a very high surface-to-volume ratio, which means that they lose heat rapidly. Eating and burning fuel almost continually is their only means of offsetting this rapid heat loss. Their diet consists of vertebrates and a mixture of insects, larvae, and other invertebrates, these latter often unearthed with the shrew's long, flexible snout. Shrews will not hesitate to tackle animals larger than themselves.

The body weight of several Asiatic *Sorex* species fluctuates by as much as 40 percent over the course of a year. In winter, in fact, the weight is considerably less than in summer. While such an adaptation would seem to handicap shrews of cold climates by excessively increasing their surface-to-volume ratio, in this case it actually gives the tiny mammals an advantage: it lessens their total winter food needs and intensifies the production of heat in the little bodies.

Arctic Hare
Lepus arcticus

Snow-white fur and black ear tips are year-round garb for the arctic hares of Ellesmere Island and northern Greenland. In warmer areas, the white coats turn brown or gray with the coming of summer. Without a backdrop of snow, white fur is a liability; in snow, the coat is an astoundingly good camouflage. In fact, these white lagomorphs sometimes behave as if they were invisible in snow, with the ironic result that people can sometimes approach the hares and capture them with their hands.

These hares are large, up to 15 pounds (7 kg) each, as compared with 8 pounds (4 kg) for a varying hare. They are not herd animals in the usual sense, but they will congregate in groups of up to one hundred, then disband. Dwarf willow and other vegetation sustains them year-round, even when they must hammer through a crust of snow with their forepaws to get at it. Evidently they get a cue, perhaps through smell, as to where the willow is buried. In the very coldest places, arctic hares have specialized jaws and teeth that help them extract every bit of precious food from narrow rock crevices and other hard-to-reach spots.

Arctic Ground Squirrel, or Arctic Souslik
Spermophilus undulatus

Arctic ground squirrels are the only arctic-tundra mammals that hibernate. In August they seek an appropriate burrowing site, usually in a well-drained bank near some body of water, where the soil is warm enough that permafrost does not form near the surface. (Generally entire communities hibernate together.) The grasses with which they line their chambers serve as effective insulation. And since ground squirrels are large rodents (as large as 25 ounces [700 g]) compared with chipmunks and mice, their surface-to-volume ratio is not particularly high. They therefore do not lose heat at an inordinately rapid rate.

In April and May snow is still on the tundra. Nevertheless, arctic ground squirrels are able to emerge from hibernation, since the last of their body fat will sustain them until fresh plant foods become available. For the next three or four months these ground squirrels forage feverishly, mate, and raise their litters of six to nine young. Then in the fall they prepare, once again, to spend nearly the next three-quarters of the year asleep.

Tundra Redback Vole
Clethrionomys sp.

Lemmings and voles, both rodents, comprise the bulk of food for tundra carnivores, and are sometimes stepped on and eaten by large herbivores such as caribou. Almost exclusively herbivorous themselves, they remain active year-round, partially by foraging under the snow in winter. Although both voles and lemmings experience population explosions and crashes about every three years, the two types of mammals differ from one another both in size (voles are smaller) and in habitat preference. Voles are rarely found on the open tundra, preferring to stay near the shrubby growth of alders and dwarf willows.

Voles are also often confused with mice, from which they differ in their smaller eyes, ears, tail, and legs. Actually their legs seem shorter than they are because their upper portions are encased in the loose skin that covers the body. Voles' bodies are a bit plumper than those of mice, and their plumpness is accentuated by their blunt noses.

There are five species of redback voles (also called bank voles). *C. rutilus* has a far-northern, nearly circumpolar range. It extends across the Soviet Union from northern Scandinavia in the Old World, and from Hudson Bay to the shores of Alaska in the New.

Norway Lemming
Lemmus lemmus

"Like lemmings to the sea"—the phrase most likely refers to the Norway lemming, famous for its so-called suicide marches to the ocean. Lemmings fan out across the tundra in great hordes, but they are no longer thought to be seeking their own death. Scientists now believe that they respond periodically to overcrowding by emigrating from their homes in large numbers. The small rodents are an easy mark for carnivores, birds of prey, and even an occasional caribou, which may crush a lemming with a hoof if it gets hungry enough. Of those that reach the ocean, many drown. A few do not emigrate, and these individuals help rebuild the population, which peaks again in three years. The entire phenomenon is a wonder, and still little understood.

The closely related collared lemmings are the only rodents whose coat changes color in winter. In their white pelage they are hard to distinguish from the albino lemmings that share their circumpolar range. In winter the claws on the collared lemming's forefeet become enlarged, forming an unusual "scoop" that evidently helps the animal burrow through crusty earth or snow. In summer the claws shrink to normal size.

* Timber Wolf, or Gray Wolf
Canis lupus

Timber wolves, once maligned for their voracious eating habits, have been hunted nearly to extinction in many parts of their range. *(C. lupus irremotus,* the northern Rocky Mountain wolf, is endangered.) To be sure, they are efficient pack hunters capable of bringing down old and infirm caribou, on whose trails they often migrate. And their lanky build and ability to move on the tips of their feet are definite assets during a chase. But wolves are hardly greedy, a human concept too often applied to animals that store extra food as a hedge against skimpy times, or that take domestic species with any regularity.

Wolves have evolved a rigidly structured social system that takes care of many needs. They stay with the same mate for life, and are exemplary parents. Both mother and father care for the young and bring food to the den. They may bring whole food, or they may digest food in the field and regurgitate it to the pups when they return to them. The latter method is a marvelous adaptation in that it allows the adults to cover long distances without the encumbrance of prey in their mouths. Should a parent die, some other member of the pack may take over the rearing of the offspring. Communal howling often precedes the hunt, and is usually accompanied by much playing and bodily contact.

Arctic Fox
Alopex lagopus

Arctic foxes have many features characteristic of cold-weather animals—short ears for minimum heat radiation; furred soles on their paws for insulation; and, in the northernmost individuals, a white winter coat that serves as camouflage in snow. More southern individuals retain a bluish-gray coat year-round. This mammal's fur is thicker, relative to the fox's size, than any other polar animal's.

At only about 14 or 15 pounds (6–7 kg), this circumpolar fox is the smallest canid predator. Its eating habits are similar to the polar bear's. In fact, the arctic fox may trail a polar bear across the sea ice and take some of the bear's leftover food (seal, preferably). It eats these leftovers warily, particularly if the bear is still nearby. (Polar bears typically eat only one-quarter to one-half of their seal kills, then either abandon them or go to sleep. The bear inadvertently feeds many other animals this way.) Lemmings are also a staple food item, as are sea birds. The arctic fox will store supplies of them in rock crevices, where larger predators cannot easily steal them.

Denning sites may be in well-drained eskers, the long ridges of debris that were left in the wake of the glaciers. In the spring, about eight weeks after the arctic fox mates, five to ten pups are born, and are usually frolicking near the mouth of the den within only a few weeks.

Wolverine, or Glutton
Gulo gulo

Wolverines are bowlegged, squat, and slow. But they will not hesitate to prey on elk or moose impeded by the snow. Or they may climb a rock and wait to spring on a passing sheep. Even bears and coyotes vanish when a wolverine appears at their kill. With such appetite and ingenuity, the glutton seems to deserve its reputation, in spite of the fact that it may subsist on nothing but berries, eggs, and insect larvae during the warmer times of year. In winter it grows a heavy coat and stiff hairs on the soles of its feet, and thus equipped for snow it does its hard-nosed hunting. In North America only the porcupine seems capable of stopping it: the porcupine's quills can pierce the wolverine's internal organs and cause death.

These are mammals of both taiga and tundra, New World and Old. They belong to the mustelid family, along with the weasels, otters, badgers, and skunks. In fact, they are often called skunk-bears because of their resemblance to both these animals. Today they are fairly rare and are almost never seen in the wild by human beings.

Stoat, or Ermine
Mustela erminea

The ermine's splendid pelage, white except for the black-tipped tail, has traditionally been worn by royalty. This is the animal's winter coat only. In warmer months the ermine takes on a mellower look as its back turns a rich tundra brown. It takes about a month for the color to change completely. The spring molt begins with a brown stripe down the center of the animal's back, from which the brown tone spreads downward. In late autumn the white pelage first appears on the abdomen, then the new tone spreads upward.

These mustelids are also known as short-tailed weasels, or in Eurasia as stoats. They are dauntless carnivores that are able to prey on rodents and rodentlike mammals because of their own small size and serpentine bodies: they can easily slip between rocks or under snow to reach small animals. They kill instantly by biting their prey in the head. Then they consume the flesh, the skin, and the bones. Quite commonly the ermine even takes over the burrow of its just-eaten prey, although not all individuals nest underground. Ermines have been known to make their homes in hollow logs, buildings, and even deserted farm equipment.

Polar Bear
Thalarctos maritimus

These loners of the pack ice, the most highly carnivorous of all bears, may never set foot on land. In the water they search year-round for seals, often swimming 20 miles (32 km) or more at a time. Despite their enormous size (up to half a ton [454 kg]), they are unusually subtle killers, stalking their prey directly on the ice or waiting for seals to pop their heads through breathing holes. They kill by delivering a single blow with the paw. Eskimos claim to have seen them cover their giveaway dark-colored nose with a forepaw while hunting on the bright ice.

Polar bears are the only arctic species in which all the individuals stay white year-round. (Arctic foxes, stoats, and varying hares change color with the seasons, and timber wolves are generally lighter-toned in the northernmost Arctic, darker-toned farther south.) Adaptations that equip them for their nomadic life style include a keen sense of smell and well-furred feet that provide insulation from cold. Polar bears tend to place their huge paws far apart, and so distribute their body weight over a large area. This is important on fragile ice floes.

Fur protects the polar bear from frigid air temperatures, and a layer of blubber serves as insulation in freezing waters. Sheets of muscle rich in blood vessels rest just under the skin on the back. These appear to be some sort of heat exchangers, though exactly how they work is not fully understood. They are particularly valuable when polar bears exercise and during warm weather, when the animals actually face the danger of overheating.

Caribou, or Reindeer
Rangifer tarandus

Armies of barren-ground caribou stream south across the tundra in autumn. They are heading for the taiga, their migration triggered by food shortages as the tundra starts to freeze. Their wide cloven hooves help them unearth lichens growing under snow; they also function as mini-snowshoes in frozen or boggy terrain. And their coat of hollow, lightweight guard hairs helps provide some of the best insulation and one of the most effective wind shields possessed by any animal, by keeping dead air trapped close to the skin.

Rutting season occupies the bucks in the fall. Fawns are born about 240 days after conception, by which time the does are well on their way back to their summering grounds on the tundra. (The bucks join them later.) Fawns are gangly, but can keep pace with the adults after only a few days.

The only true tundra caribou is the barren-ground form, the other two forms being the woodland and the mountain caribou. "Reindeer" is actually the Old World and Greenland's name for the caribou, but it is also used to refer to the domestic form of this mammal. The species is very important to Eskimos and Laplanders as a source of food and leather.

Musk Ox
Ovibos moschatus

Of all arctic land mammals, only musk oxen survive the rigorous winters of the far North without shelter of any kind. Clad in extra-heavy coats, they ride out storms in huddles—the young in the center, the adults facing outward around them. This formation helps the animals to pool body heat and provides a well-horned defense against polar bears and wolves. But it is also an invitation to hunters—so much so, in fact, that musk oxen were hunted nearly to extinction before they were finally given legal protection. Now their numbers are increasing in the New World. (Originally a circumpolar species, the musk ox has been extinct in Eurasia for thousands of years.)

Taxonomists have assigned the musk ox its own genus, since it is really neither bison nor sheep nor goat nor cow. Wholly herbivorous, it feeds in winter on meager plants swept clear of snow by the wind. Or it may paw through a crusty layer of snow to reach frozen vegetation. Winter food is usually so meager, in fact, that the animal loses weight steadily and must rely on its fat reserves to carry it into spring. In warm weather, of course, these fat deposits are rebuilt as the musk ox browses on vegetation, mostly that which flanks tundra gullies and streams.

Alpine Marmot
Marmota marmota

These are the hibernators of the Alps. Like the arctic ground squirrels of the Nearctic realm, alpine marmots undergo a yearly six- to eight-month sleep, which coincides with the length of winter. Their body temperature drops severely, and their metabolism slows down until the animals are in a comatose state in their deep winter burrows.

In summer, colonies of these marmots may be seen sunning themselves and playing in the open meadows of the Alps. (Their habitat actually ranges from clearings in the upper coniferous forests to well beyond the tree line and into the fields of year-round snow.) Grasses and flowers are their main staples, the marmots feasting in an upright position with the food held in their forepaws. If an individual senses danger, it whistles a high-pitched warning, and instantly the entire colony scurries to safety among the rocks. Sometimes curiosity compels groups of these rodents to sit up and look around instead of fleeing.

The bobac marmot, a resident of the Eurasian steppes, is a close cousin of the alpine species. Also related are the woodchuck, yellow-bellied marmot, and hoary marmot of North America.

Mountain Goat
Oreamnos americanus

Rocky Mountain goats are at home on the top of the world. Their sure-footedness derives from beautifully adapted cleft hooves. The hooves' concave soles spread open when pressure is applied, allowing the goats to grip smooth rock surfaces, even on extreme inclines.

Because of their extraordinarily efficient digestive system, these mammals are able to take advantage of the sparse vegetation that grows above the tree line. And a double layer of thick white wool keeps them warm in weather that would be intolerable to man. Only avalanches present a hazard from which there is no refuge, and mountain goats often become their victims.

Male mountain goats, or billies, maintain harems and are often antagonistic to their young. In fact, their sharp black horns can do the kids considerable damage if the nanny fails to intervene. Kids whose mothers are killed are deprived not only of protection and the invaluable survival training that the nanny provides, but are actually isolated from the herd and left to die.

Mountain goats are not true goats, but are related to the antelopes. Unlike deer and some other species, mountain goats do not easily make up losses they suffer at the hands of hunters. For this reason game authorities must monitor their numbers carefully at all times.

* Ibex
Capra ibex

Contrast the short horns of the chamois and mountain goat with the magnificent curved horns of the ibex! Older rams, of course, bear the longest horns, 2 to 3 feet (61–91 cm) or longer from tip to base. The prominent ridges on these weapons are indicators of age. Generally two or three ridges are added each year, which makes reading the ibex's age something of a guessing game.

Except during mating season, ibexes segregate themselves sexually, ewes and young usually herding together at lower altitudes than the rams. Neither sex will go below the tree line, however, since the ibex's feet are adapted to mountain precipices, not to the forest floor.

These wild goats have an interesting history. During the Pleistocene epoch they inhabited southern France, western Italy, and southern Germany. In self-defense against greedy hunters there, they gradually retreated into the Alps, where they were not given official protection until the early nineteenth century. By that time the ibexes had been hunted nearly to extinction. Since then, however, they have enjoyed a remarkable comeback and seem well established today on alpine slopes and in Italy's Gran Paradiso National Park. One subspecies, however, *C. ibex walie* of Ethiopia, is considered endangered due to habitat degradation.

Bighorn Sheep
Ovis canadensis

The handsome Rocky Mountain bighorn sheep seem invincible, with their massive bodies, curled horns, and well-adapted feet. Yet sometimes they face encroachment on their habitat; and they may also face the hunter's gun, which has caused the species to be seriously threatened. Horns are prize trophies, particularly those of the older rams (the age of bighorns can be determined by the number of horn corrugations). Unlike antlers, which are shed annually, the horns grow constantly. They serve as battle weapons during the autumn rut, the noisy head-on clashes between rams often making both rivals reel with dizziness. Fair play seems to dominate the competition, however, since only rams with equally matched horns battle each other for harems. When the young are born, it is usually on the least accessible ledge, no doubt a necessary precaution that protects the lambs from predation until they are old enough to walk.

North American bighorns are one of five types of true wild sheep in the genus *Ovis,* the other four being scattered across much of the northern hemisphere.

Chamois
Rupicapra rupicapra

Chamois occupy the same niche in the Old World that mountain goats do in the New World. Not true goats (which are classified in the genus *Capra*), these members of the *Rupicapra* genus are the most superbly adapted large herbivores of high and barren places. Chamois even seem to celebrate their particular place in the world, since they perform lovely "dances" on their cloven feet, and slide down alpine slopes on their stomachs without hesitation. One can only suppose that they expend such energy simply for the joy of playing.

The feeding habits of these ungulates are more varied than those of the ibex. Chamois are comfortable feeding in forests, in snow, and in alpine-tundra meadows, and they move from higher to lower altitudes with the cold. Usually they will not pass up a meal of mushrooms, and they enjoy a variety of fresh shoots. Even flowers and herbs are part of their diet.

Unlike the mountain goat, the chamois does not exclude abandoned kids from the herd and leave them to die. Rather, the herd adopts them and, by doing so, no doubt contributes to the preservation of the species.

8

The Ocean

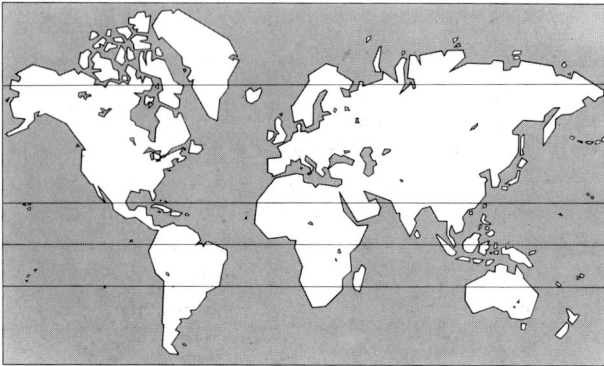

Mammals that are highly adapted for aquatic life are usually classified in one of three orders: the cetaceans (whales, dolphins, and porpoises); the pinnipeds (seals, walruses, and related forms); and the sirenians (dugongs and manatees). The sea otter, a member of the carnivore order, is also considered an aquatic species. What sets these animals apart from beavers, muskrats, platypuses, and other mammals that spend a large part of their time in the water is the degree to which they are specialized for their liquid environment.

As a group, marine mammals differ tremendously from the terrestrial mammals, especially the oceangoing cetaceans, which, like the sirenians, never leave the water of their own volition. They are huge in comparison with land mammals; they swim instead of walk, and have appendages modified for that purpose; some of them bear their young tail-first instead of head-first; many have blubber instead of a coating of hair; some of them orient themselves and communicate with each other through means quite unlike those used by most other mammals; and several even feed in a manner that is totally different from the way land mammals feed. These are such major differences that marine researchers and biologists have found that their usual assumptions about mammals are out of place and have been forced to try to understand the brain of the ocean-dwelling species on its own very special terms.

Size is the most striking difference between land and marine mammals. The blue whale, *Balaenoptera musculus,* has been known to attain a weight of 49.5 tons (45 m t), which easily qualifies it as the largest animal ever to have lived on the planet. The fact that it can outweigh even the biggest dinosaurs by more than two to one is spectacular yet not surprising, given the differences in the animals' environments: dinosaurs, like all other land mammals, had to support their body weight on their own appendages, while marine mammals have their weight supported for them by the ocean.

The very shape of marine mammals indicates their efficient method of locomotion. Their torpedolike bodies greatly minimize water resistance during swimming. The hairless skin of the cetaceans streamlines them further. (Actually, some whales do have hair on their heads, but there is so little of it that is has no effect on movement.) The pinnipeds, of course, are covered with fur, but nevertheless are remarkably sleek and agile in the water. The members of one pinniped family, the Phocidae, have evolved such a streamlined body form that external ears and reproductive parts have been completely eliminated from their bulletlike shape.

Cetaceans propel themselves by an up-and-down movement of their horizontal tail flukes. Unlike fishes' tails, these flukes have no skeletal support, but are made up entirely of tendons and gristle. In comparison with the rest of a whale's body, the tail flukes are small. Yet they displace a great deal of water with each thrust and can move a blue whale through the sea at a rate of 20 knots.

Paradoxically, the smaller whales such as dolphins, which would seem to have much more limited power, can typically outrun a blue whale. The difference is in the structure and composition of the dolphin's skin, which allows water to flow over it so as to reduce up to 90 percent of the potential drag. The skin hangs loosely on the dolphin's body, and its outer layer is composed of a spongelike material. Consequently it moves with the water flowing past it instead of remaining rigid and opposing the direction of flow.

Preceding pages: Northern fur seals, Arctic Ocean

Cetaceans use their paddlelike forelimbs primarily for balance and steering. This is not necessarily the case with the pinnipeds. While some pinnipeds propel themselves by sculling with the hind flippers, others get their stroking power from the forelimbs, and the muscular forward portion of their bodies reflects this difference.

All mammals breathe oxygen, of course—even those that inhabit the sea. Cetaceans are specially equipped for efficient surfacing and breathing, since their nostril, or blowhole, is on the top of the body. This means that they need expose only a minimal portion of their backs to the surface in order to draw a breath. Usually a whale breathes every 15 to 60 seconds, and may sleep just at the water's surface.

What seem to landlubbers to be incredibly impressive respiratory feats are part of many whales' normal activities. For example, the sperm whale, *Physeter catodon*, may hold its breath and "sound," or dive vertically, to a depth of well over 3,000 feet (915 m), and it may stay submerged for up to an hour. This extraordinary breathing is possible partly because whales' lungs function very efficiently. While they actually have a surprisingly small oxygen capacity, they can be nearly filled each time a whale inhales. In addition, only 9 percent of inhaled oxygen is held in a whale's lungs, as compared with 34 percent in human beings. In whales 82 percent of the oxygen is distributed to the muscles and blood, another 9 percent to other parts of the body. This is in comparison with only 54 percent distributed to human muscles and blood, 12 percent to other parts of the body. Therefore, rather than having to rely heavily on oxygen that has just been inhaled, as human beings must, the sperm whale is able to draw on its oxygen reserves to sustain it over long periods. And because the blood's circulation rate is slowed considerably during dives, oxygen is not used up by the tissues as fast as it is at other times. (Blood flow to the brain is not really slowed. If it were, the whale would lose consciousness and soon die.)

When a whale surfaces, its blowhole opens automatically. Then the animal exhales and inhales until it repays the "oxygen debt" it has accumulated. Usually the whole process takes very little time, thanks to the extremely efficient lungs.

What observers see rising from a surfaced whale depends on the species. Whales known as rorquals exhale a single plume straight into the air; right whales a double plume, the two sides of which form a V; and so forth. The contents of the exhalation, or blow, are still something of a mystery. Certainly the blow contains condensed water vapor from the lungs. In addition, it probably contains some ocean water that has gotten into the animal's air passages. It may also contain an oily foam. This substance, it has been hypothesized, absorbs nitrogen during dives and may help prevent whales from getting the bends, the painful condition brought on by too much nitrogen being accumulated, under pressure, in the blood.

Blubber insulates the cetaceans against cold ocean temperatures; it also houses the numerous vessels that circulate blood close to the skin surface for cooling the body when it heats. The blubber is located between the muscles and the skin, and in some of the whales it may be several inches thick. Pinnipeds regulate their body temperature somewhat differently, since all of them except the walrus have fur as well as blubber. While some of them rely exclusively on the blubber to keep them warm in water, others have a double layer of fur that traps air bubbles very efficiently in water, and helps prevent heat loss. In addition, seals of the Phocidae family have a high rate of metabolism and a low rate of blood flow through the flippers, and both of these adaptations help conserve body heat. The sea otter, which has no blubber at all, relies on its lush, dense fur to keep it warm. If the fur becomes heavily soiled or oily, as it does in oil spills, it loses its insulating properties. Mother sea otters, apparently aware of this danger, groom their offspring regularly.

Relatively little is known about the courtship, mating, birth, and parenting behavior of most marine mammals, especially the sirenians. It is known that both the cetaceans and the sirenians mate, bear their young, and nurse in the water, while most pinnipeds carry out these functions on dry land. Sea otters court and mate in the water, but the females go ashore to give birth.

Perhaps the most distinctive aspect of cetacean birth is that the offspring are born tail-first, rather than head-first like most other mammals. If the head did not leave the mother's body last, the newborn would not be able to reach the surface in time to breathe, and would therefore drown. If, when its head at last emerges, the infant needs a nudge to the surface to breathe, its mother is free to help it, usually along with a "midwife" or "aunt" or two. First breaths fill the newborn's lungs with oxygen and also help to stabilize the animal's buoyancy, which is important since the scant supply of blubber in whale newborns limits flotation somewhat. The offspring nurse by merely touching the mother's nipple, which squirts large quantities of rich milk into the waiting mouth. This arrangement frees both infant and mother from long sessions of underwater suckling. (The milk of the blue whale is so rich that the infants gain weight at the extraordinary rate of 10 pounds [4.5 kg] an hour!)

Pinnipeds undergo an altogether different type of reproduction. Many of them have strongly developed homing instincts, which guide them each year to specific breeding grounds. There they give birth amidst others of their kind. The females usually mate again while they are still nurturing the newborns. Because of delayed implantation, however, development of the embryo stops when it has reached a certain stage, and birth does not occur until the following year.

There are also significant differences in terrestrial and marine mammals' sensory perceptions. In the ocean, hearing is often more useful than sight, especially among the deep-diving cetaceans. Ocean sediment, lack of light at lower levels, and, in sperm whales at least, a head configuration that precludes stereoscopic vision all serve to make seeing underwater difficult. Perhaps more important, sound waves travel about five times faster in ocean water than in air. Therefore using underwater sound for navigation purposes would seem to make as much sense

for marine mammals as it has for man, who has used sonar in submarine work for quite some time.

Bottle-nosed dolphins and some of the other cetaceans emit sounds that have been variously described as whistles, clicks, whines, smacks, barks, moans, and mews. Some of these sounds no doubt constitute a language of their own, and whales are thought to be extremely responsive to each other's communications. Others, which serve to help the whales navigate, are part of a system of echolocation signals not unlike those used by bats in the air: certain sounds strike objects in a path and bounce signals back to a receiver. The sound receptor and the brain seem capable of sound discrimination far surpassing that of other mammals. The brain's overall structure is proportionately wider than that of terrestrial mammals (including human beings), and this is thought to reflect the cetaceans' sensitivity to sound, especially since their acoustic nerve and attendant sound-receiving mechanisms are very large indeed.

Cetaceans, of course, use their large, complex brains

Manatee, Florida

for far more than sound transmission and reception, although exactly what they use them for and how they use them are still unknown. The cerebral cortex is the section of the brain that scientists use to measure the evolutionary progress of a species, and in whales it is both intricately convoluted and proportionately very big. Furthermore, it seems to have developed along altogether different lines from those of all other mammal orders except one—the primates. Whatever superior intelligence has been demonstrated or assumed in chimpanzees, for instance, seems implicit in cetaceans as well.

Since the manipulation tests appropriate for primates and other terrestrial animals are somewhat impractical for measuring the brain power of whales, perhaps scientists will look even more closely at the whale's eager playfulness as a clue to its intelligence. This is a standard measure of intelligence in all mammals, and some whales are thought to relish performing tricks, and even to welcome human affection and approval while still safeguarding their own kind. What makes the bleak future of cetaceans so tragic and ironic is the fact that whaling threatens to wipe out entire marine species for products as dispensable as cosmetics and pet food just when scientists are beginning to tap the whale's awareness and to glimpse what whales might really have to contribute—a better understanding of the bonds that link all living things.

All living cetaceans are divided into two suborders according, primarily, to the manner in which they feed. Only ten species comprise the Mysticeti (baleen or whalebone whales). They are much larger than the members of the other suborder, the Odontoceti, or toothed whales.

Baleens have been called the grazers of the sea, since they feed primarily on plankton, the small free-floating organisms such as krill that populate the oceans. They open their mouths and filter ocean water through huge baleen plates that hang from their upper jaws. Then they bring the upper and lower jaws together, forcing out the water with their tongues. What remains in their mouths are scores of tiny but nourishing living things.

The toothed whales are the much more numerous group, totaling about sixty-seven species. They feed like the carnivores in that they prey on relatively large single individuals such as seals, squids, and fishes, rather than taking multitudes of tiny ones each time they swallow. Odontoceti teeth are undifferentiated from one another, all of them generally sharp and conical. Although these whales use their teeth to capture prey, they do not use them to chew their food, but leave the entire job of digestion to their several stomachs. Toothed whales have a single blowhole, as compared with the baleens' double one. The toothed whales often dive for their food instead of feeding entirely in the upper oceans, as the Mysticeti do on plankton. They generally fill many more ecological niches than the baleens do, and today a few of them may even be found in fresh water instead of open seas.

Whereas cetaceans have lost their two functional rear limbs, the pinnipeds (which means "fin-footed") still have theirs. This gives them the ability to maneuver on land (although in some cases with very little grace) and thus makes it possible for them to reproduce ashore.

Like the cetacean order, the pinniped order is subdivided. The true seals, or phocids, are the most streamlined of the three subgroups, having no external ears or reproductive organs to slow them down in the water. The fur seals and sea lions, known collectively as the otariids, do have visible ear flaps and external sexual parts, and can move about on land relatively easily, since their hind flippers are not set permanently facing backward as the true seals' are. Otariids and phocids differ also in their method of movement in the ocean, the former getting their power from the front flippers, the latter from the rear ones. Walruses, also included in the pinniped order, occupy a subdivision of their own.

The sirenians, the third order of marine mammals, has only four living species, three of which are in the same genus. These are the only totally herbivorous mammals in the sea. They are named for the sirens, who, according to classical mythology, seduced sailors with their songs.

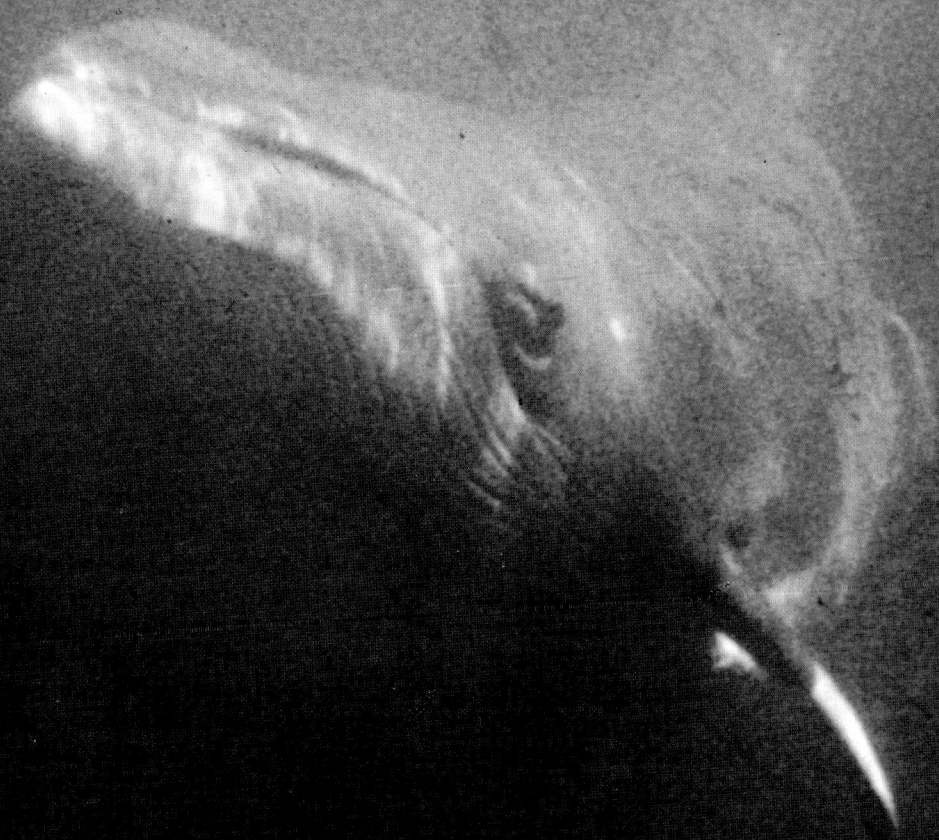

Minke Whale
Balaenoptera acutorostrata

Only 29.5 feet (9 m) long, minke whales are the smallest members of a group of baleens known as the rorquals. The word "rorqual" comes from the Norwegian term for "tube whale," and refers to the obvious pleats, or folds, of skin that run along the underside of the whale's body from the jaw to the chest region. The function of the folds is unclear, but they may contribute to the rorquals' impressive swimming speeds—under pressure, these whales can exceed 12 knots, which is faster than whales without the skin folds can move. In spite of their speed, the rorquals (which also include the blue whale, the fin, the sei, the humpback, and the Bryde's whales) constitute the main catch of modern whalers.

Rorquals typically feed in the cold waters of the northern latitudes, then decrease their intake or cease feeding altogether when they move to the warmer waters of the middle latitudes for breeding. Minke-whale migrations have been plotted less thoroughly than most of the larger rorquals'. It is known, however, that minkes are widely distributed and even penetrate far south into Antarctic waters, where they have been seen popping open the ice with their noses in an effort to create breathing holes.

California Gray Whale
Eschrichtius robustus

Each year the California gray whale migrates a total of between 10,000 and 16,000 miles (16,090–25,744 km) round-trip, from the Bering Sea to California and the lagoons off the Baja peninsula. This is the longest known mammal migration, and it is carried out for the express purpose of mating and giving birth in warm, shallow water. The females, which weigh between 27.5 and 38.5 tons (25–35 m t), often seek out the saltiest water, where they can deliver their 1.6-ton (1.5-m-t) young with maximum support from the sea. The newborns can neither float nor swim and must be held on the surface of the water in order to breathe. The mother simply carries the infant on her back to keep it on the surface.

When California gray whales mate, the female lies on her back in the water next to the male, and the two turn toward each other on their sides as they try to copulate. Often a second male spreads himself across the coupled pair and helps them hold their position. Synchronized breathing is a problem, and the act of mating may be attempted every few minutes for well over an hour before it is successfully completed.

Gray whales were nearly wiped out by whalers who killed them at their breeding places before protective legislation was passed. Today they are recovering steadily.

* Black Right Whale
Eubalaena glacialis

When whaling was done from open boats with hand-operated harpoons, right whales were the "right whales" to take, since they were relatively slow-moving and did not sink after they were dead. In fact, they constituted the main catch of whalers for so long that by the early twentieth century there were almost none of them left. Since that time right whales have been protected; yet one of the three species is still so scarce that it is practically unknown today. This is the bowhead, or Greenland, right whale, which has about three hundred baleen plates that measure up to 14 feet (4 m) each in length. The longest baleen plates known, they are used to trap plankton as the whale skims across the surface of arctic waters with its mouth open. This is only one way of foraging, the other two being for the baleen whale to gulp huge mouthfuls of food (as the rorquals do) and somehow to feed on the bottom (as gray whales are thought to do).

The black right whale is a somewhat smaller version of the bowhead. Although still far from plentiful, black right whales have been sighted in the Atlantic with some frequency during the past several years, and there is some hope that they will recover their former numbers.

Sperm Whale, or Cachalot
Physeter catodon

The sperm whale is the largest of the Odontoceti (males reach 59 feet [18 m]) and the only one now hunted commercially in sizable numbers—fifteen thousand are slaughtered each year. It is essentially a warm-water whale, and the only individuals found in polar seas are the solitary bulls. Since sperm whales are polygamous and the bulls maintain harems, these solitary males may be rejected suitors, but no one knows for sure. The species is highly gregarious, with as many as several hundred whales often traveling together. They tend to get stranded easily, since they seem to be willing to follow a herd leader blindly.

Spermaceti, from which this whale gets its common name, is a substance housed in the animal's melon, or rounded snout. Whalers used to think it was the animal's seminal fluid ("spermaceti" means "seed of the whale"), but it is now known to help regulate buoyancy. Another substance, ambergris, is still in demand as a fragrance stabilizer in perfumes. It is found in the sperm whale's intestines and, like spermaceti, is produced only by this species. It may be some sort of by-product of the beaks of squids, on which sperm whales feed quite heavily.

An interesting feature of the sperm whale is its unusual skull, the most asymmetrical of any animal known.

Narwhal
Monodon monoceros

In addition to being the northernmost of all whales (it rarely ventures below 65° N latitude), the narwhal may have given rise to the legend of the unicorn: whalers who sold narwhals' spiraled tusks to landlubbers usually did not tell their customers where they had gotten them. Until the end of the nineteenth century both the ivory yielded by the tusks and the high-quality oil from narwhal blubber were sought after, but today there is little demand for either. If the narwhal's tusk serves some particular purpose other than that of a secondary sex characteristic, no one seems to have determined what it is.

The narwhal and the beluga are the only two members of the Monodontidae family, the white whales. The two species are so closely related that their young are almost impossible to distinguish. As the beluga ages, its white skin gradually becomes yellow; because of this, and because it was the first whale ever to make sounds that were quite audible to whalers, the beluga was once known as the sea canary.

Killer Whale
Orcinus orca

Killer whales have a longstanding, and probably well-deserved, reputation for ferocity. They are the only cetaceans to eat warm-blooded vertebrates, other cetaceans and seals being among their favorite foods. Their attacks are often well organized, with the older males keeping watch at the shoreline while the other individuals hunt in shallow waters. (The killer whale is a highly social animal with a well-structured society.)

The killer whale's ferocity, however, does not seem to extend to human beings. Ever since the first killer whale was captured and publicly exhibited, in 1965, the species has proved nothing but quick to learn and cooperative. It responds to music, will allow a trainer to gambol with it in the water (and even to put his head in its open mouth), and is thought unique because its trust can be earned through simple human contact instead of rewards of food. If some sort of language communication is ever to be established between human beings and other animals, it may be through either this species or the bottlenosed dolphin.

▲ Narwhals ▼ Killer whales

▲ Amazon (freshwater) dolphin ▼ Bottle-nosed dolphins

Freshwater Dolphins
Inia geoffrensis and others

Not all dolphins live in the sea. Scientists have classified the four genera of freshwater dolphins in the Platanistidae family—*Inia, Platanista, Lipotes,* and *Pontoporia.* They are found in the rivers of India, China, and South America, and one of the four is thought to migrate in the coastal waters of South America as well as to inhabit estuaries. The main difference in conformation between these and seagoing dolphins is the freshwater species' long snout, which is used to probe about on river bottoms for food such as fish and shrimp.

The Amazon dolphin, *Inia geoffrensis,* has been the subject of limited scientific observation, but freshwater dolphins as a whole are still not very well understood. In captivity, *I. geoffrensis* seems just as solicitous of the human touch as the saltwater species, but it apparently does not make the dramatic leaps that its seagoing relatives do. Its skin is very soft, and the skin around the blowhole tends to be darker than elsewhere. The concentration of pigments there may help prevent sunburn when the animal surfaces to breathe in clear waters.

Bottle-nosed Dolphin
Tursiops truncatus

By now the dolphin is well known for its remarkable performances in marine aquariums. Its body is agile and it learns its amusing tricks easily. What is perhaps less well known is the astonishing research that is being done on captive dolphins. Much of it has to do with their echolocation and communication systems, both of which appear to be highly sophisticated. In fact, some scientists suspect that they have a complete language. It is known that they can imitate certain human sounds much as parrots do, and published accounts of work with these animals offer hope that a real sense of understanding, based on a language that is mutually comprehended by dolphins and human beings, will one day be a reality.

The actual mechanisms responsible for dolphins' sound production and reception are still not clear to us. Cetaceans have no vocal cords; dolphins seem to make sounds by vibrating some sort of structure near the blowhole. Sounds that bounce back to the dolphin are picked up by its sensitive ear.

T. truncatus breeds in spring, and the females undergo a gestation period of some eleven to twelve months. Calves seem to be on individual weaning schedules—some of them are fed squid at only six months, while others rely on milk for up to a year and a half.

Harbor Porpoise and Spinning Dolphin
Phocoena phocoena and *Stenella longirostris*

The terms "dolphin" and "porpoise" are often used interchangeably, and are not true scientific designations. As a rule, dolphins are those whales that have distinct beaks, sickle-shaped back fins, and pointed teeth. Porpoises are whales with rounded heads that do not have well-differentiated beaks. They have triangular back fins, and their teeth are truncated, not pointed.

Whereas most dolphins and porpoises are exuberant in the water, leaping about and riding the bow waves of ships, the harbor porpoise is rather retiring. Usually it ignores vessels speeding past it. It is at home in coastal waters and the mouths of large rivers, and has been known to swim up rivers. The spinning dolphin, or spinner, is as playful as the harbor porpoise is reserved. Its common name is derived from its habit of leaping out of the ocean and spinning before it dips back into the sea. Other cetaceans have been known to imitate these spins.

Although the nature of the bond that exists between tuna and dolphins is still a mystery to science, the tuna industry has increased its catches enormously by letting dolphins lead fishing boats to the tuna schools with which they swim. Because of this practice, three species of dolphins have been injured or killed in large numbers (estimates range between three and six million). *Stenella attenuata* (the spotted dolphin), *S. longirostris* (the spinner), and *Delphinus delphis* (the common dolphin) are trapped, along with the tuna, by purse seines, specialized nets used by tuna fishermen. While theoretically these dolphins can be spared if handled carefully, all too often they become prey for waiting sharks, or they panic and behave in unpredictable ways that bring about their death. Research is now being done to explore the bond itself and to develop fishing methods that will allow maximum tuna catches while still guaranteeing the dolphins' safety.

Sea Otter
Enhydra lutris

Sea otters are known for their devotion to their young, and have been said to weep when their offspring become separated from them or die. The mother keeps the young close to her own body for much of its first fifteen months, and is affectionate as well as efficient in caring for it. She may place it in a crook in her arm when she dives, and she often carries it on her chest when she floats. Frequently she will anchor it in seaweed as she feeds and may even wrap herself in the seaweed to prevent storms from carrying her too far out to sea.

This captivating mammal is a tool-user. It uses rocks to smash mussel, clam, and other shells that it holds on its chest with its forepaws as it floats. It even favors its right forepaw over its left, an instance of "handedness" that is very unusual in nonhuman animals.

The lush fur of this mustelid nearly led to the animals' demise at one time. But today the sea otter is protected, and is enjoying a remarkable comeback. The waters around the Aleutian Islands are home for most sea otters today.

Harbor Seal
Phoca vitulina

Harbor seals move so poorly on land that it is surprising that they come ashore as often as they do. In the water, however, they are marvelous to watch as they scull with their hind flippers. Like some cetaceans, they echolocate; and they see extremely well for a marine species. This is due to a flattened cornea, which permits them to focus their eyes underwater. The eyes "weep" almost continually, perhaps in an effort to rid the body of ingested sea salt by excreting it through the tear ducts. Only after such purification would sea water seem to be useful to the harbor seal's system.

This pinniped's habit of swallowing stones has puzzled scientists for quite some time. It may be that the stones provide needed weight when the seal dives for the various fishes and shellfish that make up its diet. It is thought that the harbor seal cannot float on its back and nap in the water as some other seals do.

Harbor seals are widespread throughout the coastal areas of the northern hemisphere, and they sometimes venture into fresh water.

Elephant Seal
Mirounga sp.

The male elephant seal, the largest of all pinnipeds (up to 4 tons [3.5 m t]) has an interesting proboscis, which resembles a horse's nose in texture when relaxed. During breeding season, however, it becomes inflated to nearly twice its usual size. Its function is unclear; perhaps it acts as a resonator for the animal's trumpetlike sounds.

When the adults haul onto land to mate and give birth in the spring, they are extremely fat from their winter diet of squid. A season of defending harems and suckling young, however, leaves no time for feeding, and elephant seals lose an astonishing amount of weight before they return to the sea. The mother seals' milk is one of the richest substances known, usually composed of 50–80 percent butterfat, as compared with a maximum of 10 percent in cow's milk. Not surprisingly, the growth of the single pup is as rapid as the mother's weight loss.

One of the most fascinating adaptations of these and other pinnipeds is their seeming oblivion to pain. During mating, the males bite and hold the females' necks, leaving dozens of tooth marks. And the bulls' bodies are literally covered with battle scars from repeated encounters with rivals attempting to take over harems. The animals are either insensitive or somewhat stoic about the injuries they receive.

Northern Fur Seal
Callorhinus ursinus

From their wintering grounds off Alaska, populations of male northern fur seals begin arriving in early spring at the Pribilof Islands, the Commander Islands, and Robben Island. By the time the females join them, around mid-June, the bulls have engaged in numerous battles for territories. During this time and right through mating with the harem members, they do not leave their chosen spots even to eat, and they are emaciated when they return to the water in midsummer.

The females, which migrate from areas far to the south, bear their single pups soon after they reach the breeding ground. They have been pregnant for almost a year; because of the long gestation period, the pups are furred and have their eyes open when they are born. For their first week of life the newborns are carefully nurtured by their mothers. After this time, the females alternate nursing with diving for their own food. They also mate during this period, and thus begin anew the cycle that will culminate in the following year's offspring.

The pelts of these seals are so lush that at one time the species was almost wiped out by hunters. In the early twentieth century an international pact signed by Canada, Russia, Japan, and the United States limited hunting, and today the population is up to at least 1.5 million.

*California Sea Lion
Zalophus californianus

As a rule, sea lions inhabit warm waters, to which they are adapted by having relatively little body insulation. They have blubber but no underfur beneath their coat of hair. For this reason they are not attractive to furriers, and have been spared the kind of wholesale slaughter that certain fur seals have had to face.

California sea lions are among the most highly territorial of all mammals. Bulls stake out territories that typically extend to the water's edge, and defend them through chest-to-chest combat, biting, and other tactics. With their shaggy necks and manes, the enormous bulls resemble lions—thus the common name. They may reach a weight of 617 pounds (280 kg), and they assemble harems of up to fourteen or fifteen females each.

Females and pups are gregarious and seem insecure when not surrounded by others of their general age and size. On land the pups' play consists of mock fighting and territorial games; in the water, they may leap over each other's bodies. Sea lions' barks and honks, along with their ability to learn tricks, are responsible for their popularity as circus "seals." The Japanese sea lion, *Z. californianus japonicus,* is endangered.

Walrus
Odobenus rosmarus

The walrus's family name, Odobenidae, means "those that walk with their teeth," and it refers to the animal's means of hoisting itself onto ice floes by hooking its tusks over their edges. The tusks are also used to break open breathing holes in the arctic ice. And they are weapons that can inflict serious wounds in rival bulls. Walruses are highly thigmotactic—that is, they maintain close physical contact with one another—and the tusks may play a part in this aspect of their behavior as well. Although the primary purpose of huddling is probably to share body heat (walruses have no fur), the closely packed bodies may also provide resting places for the tusks, which make it awkward for walruses to lie in a prone position. The tusks' most important function is that of dislodging bivalves from the ocean floor. The clams on which the walruses feed usually bury themselves in mud or among rocks, and the walruses must force them out.

An inflatable pharynx provides buoyancy during naps in the water, but it is also responsible for many a walrus's downfall, since the bell-like sound it produces often guides hunters to their prey. Walrus meat and skins have long supplied arctic peoples (and others) with many of their needs. But hunting is restricted now that walrus populations have declined.

Dugongs and Manatees
Dugong sp. and *Trichechus sp.*

Dugongs and manatees make up the rather bizarre-looking Sirenian order. They inhabit the eastern and western hemispheres, respectively, and have generally similar habits. Slow-moving and rather harmless (though curious), they feed exclusively on the vegetation of coastal waters. They submerge to forage, then come up to breathe every few minutes. Flippers are sometimes used for walking on shallow bottoms.

The main differences between the two types of animals have to do with physical details such as well-notched tail fins in dugongs as compared with rounded ones in manatees. Dugongs' upper lips are not so obviously cleft as the manatees', and some dentition differences exist.

Florida manatees, *Trichechus manatus,* have been hunted excessively, and are often the victims of accidents in which they become caught in motorboat propellers.

Photo Credits

BC—Bruce Coleman, Inc.   TO—Tierbilder Okapia

Front cover: Marvin Newman.
Back cover: Bob Campbell, BC.
Title page: Joe Branney, Amwest.

1. The Tropical Rain Forest

24-25: G. D. Plage, BC; 28: John Shaw, BC; 29: BC; 30: TO; 31: D. Schmidt, TO; 32: Nina Leen; 34: (top) Frank W. Lane; (btm) P. Nurs, TO; 35: C. M. Hladek, Jacana; 36: C. D. Dodge & D. R. Thompson, BC; 38: H. Root, TO; 39: A. Root, TO; 40: J. H. Carmichael, BC; 41: BC; 43: (top) TO; (btm) L. L. Rue; 45: Bob Campbell, BC; 46: P. Nurs, TO; 47: A. Root, TO; 48: Norman Tomalin, BC; 49: Kenneth W. Fink, BC; 50: Varin-Visage, Jacana; 51: TO; 53: (top) G. D. Plage, BC; (btm) Simon Trevor, BC; 55: (top) R. Hofmann, TO; (btm) P. Nurs, TO; 57: (top) Nina Leen; (btm) BC; 58: (top left & btm): Douglas Fisher, Frank W. Lane Agency; (top rt) TO; 61: BC; 62: BC; 63: Kenneth W. Fink, BC; 64-65: (top & btm) Erwin A. Bauer; 66: A. Root, TO; 67: TO; 68: D. & R. Sullivan, BC; 69: Fritz Prenzel, BC; 70: (top & btm) A. Root, TO; 71: Warren Garst, Tom Stack & Assocs.

2. The Temperate Deciduous Forest

72-73: Thase Daniel, BC; 76-77: John Shaw, BC; 78: L. L. Rue III; 79: L. L. Rue III; 80: L. Riley, BC; 81: (left) Erwin A Bauer; (rt) BC; 82: L. L. Rue III; 83: L. L. Rue III; 85: (top) H. Reinhard, BC; (btm) L. L. Rue III, BC; 86: Charles G. Summers, Jr., Amwest; 87: Erwin A. Bauer; 88: (top) J. Markham, BC; (btm) H. Reinhard, BC; 94: H. Reinhard, TO; 95: TO; 96: Jane Burton, BC; 97: H. Reinhard, BC; 98: (top) TO; (btm) H. Reinhard, TO.

3. The Coniferous Forest

100-101: Jonathan T. Wright, BC; 104: Gene Ahrens, BC; 107: Frank W. Lane, BC; 180: S. C. Bisserot, BC; 109: H. Reinhard, BC; 110: Schrempp, TO; 111: TO; 112: (left) Erwin A. Bauer; (rt) W. Ruth, BC; 113: Jen & Des Bartlett, BC; 114: L. L. Rue, Amwest; 115: (top & btm) L. L. Rue III; 116: L. L. Rue III, BC; 117: Joe Branney, Amwest; 118: L. L. Rue; 119: Erwin A. Bauer.

4. The Tropical Grasslands

120-121: Jack Couffer, BC; 124: Francisco Erize, BC; 125: Erwin A. Bauer; 126: (left) Arthur Christiansen, Frank W. Lane Agency; (rt) A. Root, TO; 127: L. L. Rue III, Amwest; 128: Frank W. Lane, BC; 129: (top) Lee Lyon, BC; (btm) Clem Haagner, BC; 130: John Kermill, Frank W. Lane Agency; 131: L. L. Rue, Amwest; 132: Simon Trevor, BC; 133: N. Myers, BC; 135: Erwin A. Bauer; 136: (top) Erwin A. Bauer; (btm) N. Myers, BC; 137: Erwin A. Bauer; 138: L. L. Rue III; 139: Kenneth W. Fink, BC; 140: Bob Campbell, BC; 141: H. Reinhard, BC; 142: Erwin A. Bauer; 143: L. L. Rue III; 144: Warren Garst, Tom Stack & Assocs.; 145: Erwin A. Bauer; 146: Australian Information Service; 148: Keystone Press Agency; 149: L. L. T. Rhodes; 151: Erwin A. Bauer.

5. The Temperate Grasslands

152-153: Lynn M. Stone, BC; 156-157: Dr. E. R. Degginger; 159: Frank W. Lane; 160: P. Nurs, TO; 161: TO; 162: TO; 163: J. Foott, BC; 165: (top) Charles G. Summers, Jr., Amwest; (btm) John S. Flannery, BC; 166: L. L. Rue III; 167: BC; 168: (top) L. L. Rue, BC; (btm) BC; 169: Frank W. Lane; 170: Francisco Erize, BC; 171: L. L. Rue, BC.

6. The Desert

172-173: J. Burt, BC; 176-177: (top) Kenneth W. Fink, BC; (btm) P. Ward, BC; 179: (top) J. Dragesso & F. Pette, Jacana; (btm) Jane Burton, BC; 180: BC; 181: (top) Thase Daniel; (btm) J. Markham, BC; 182: L. L. Rue III; 183: TO; 184: Robert P. Carr, BC; 185: BC; 186: J. Wallis, BC; 187: Australian Information Service; 188: (top) J. Wallis, BC; (btm) BC; 189: Australian News Service; 190: S. C. Bisserot, BC; 191: (top) TO; (btm) Bayer, TO; 192: N. W. Lobanow, TO; 193: Peter Carr-Locke, BC; 194: Roger W. Barbour, Univ. of Kentucky; 195: L. L. Rue III; 197: (top) H. Flygare, BC; (btm) Karl H. Maslowski; 198: TO; 199: L. L. Rue, Amwest.

7. The Tundra

200-201: H. Reinhard, BC; 204: Kenneth W. Fink, BC; 206: Fred Bruemmer; 207: L. L. Rue, Amwest; 208: Erwin A. Bauer; 209: Robert L. Dunne, BC; 210: Fred Bruemmer; 212: Francisco Erize, BC; 213: Fred Bruemmer; 214: Amwest; 215: P. Nurs, TO; 216: Charles G. Summers, Jr., Amwest; 217: Frank W. Lane.

8. The Ocean

218-219: D. K. O'Dell, Sea Library; 223: Patricia Caulfield, Animals, Animals; 224: S. Williamson, BC; 225: Franz J. Camenzia, BC; 227: (top) Fred Bruemmer; (btm) BC; 228: (top) Jen & Des Bartlett, BC; (btm) Erwin A. Bauer; 230: L. L. Rue, Amwest; 231: TO; 232: Erwin A. Bauer; 233: L. L. Rue, Amwest.

Bibliography

Allen, Durward L., *The Life of Prairies and Plains.* Our Living World of Nature Series. New York, McGraw-Hill, 1967.

Banfield, A. W. F., *The Mammals of Canada.* Toronto, University of Toronto Press, 1974.

Bourliere, François, *The Land and Wildlife of Eurasia,* rev. ed. Life Nature Library. New York, Time-Life Books, 1967.

———, *Mammals of the World, Their Life and Habits.* New York, Knopf, 1955.

———, *The Natural History of Mammals,* 3rd ed. New York, Knopf, 1964.

Bueler, Lois E., *Wild Dogs of the World.* New York, Stein & Day, 1973.

Burton, Maurice, and Burton, Robert, eds., *The International Wildlife Encyclopedia,* Vols. 1–20. New York, Marshall Cavendish, 1969–70.

Cahalane, Victor H., *Mammals of North America.* New York, Macmillan, 1961.

Carrington, Richard, *The Mammals.* Life Nature Library. New York, Time-Life Books, 1963.

Corbet, G. B., *The Terrestrial Mammals of Western Europe.* Chester Springs, Pa., Dufour, 1966.

Cox, C. Barry, Healey, Ian N., and Moore, Peter D., *Biogeography, an Ecological and Evolutionary Approach,* 2nd ed. New York, Halsted Press, 1976.

Dittmer, Howard J., *Modern Plant Biology.* New York, Van Nostrand Reinhold, 1972.

Durrell, Gerald, *A Bevy of Beasts.* New York, Simon & Schuster, 1973.

Emmel, Thomas C., *Worlds Within Worlds, an Introduction to Biology.* New York, Harcourt Brace Jovanovich, 1977.

Ewer, R. F., *The Carnivores.* Ithaca, N.Y., Cornell University Press, 1973.

Farb, Peter, *Ecology.* Life Nature Library. New York, Time-Life Books, 1963.

Grzimek, Bernhard, ed. *Grzimek's Animal Life Encyclopedia,* Vols. 10–13. New York, Van Nostrand Reinhold, 1972–75.

———, *Grzimek's Encyclopedia of Evolution.* New York, Van Nostrand Reinhold, 1976.

Guggisberg, C. A. W., *Wild Cats of the World.* New York, Taplinger, 1975.

Hylander, Clarence J., *Wildlife Communities, from the Tundra to the Tropics in North America.* Boston, Houghton Mifflin, 1966.

Kurten, Bjorn, *The Age of Mammals.* New York, Columbia University Press, 1972.

Leopold, A. Starker, *The Desert.* Life Nature Library. New York, Time-Life Books, 1962.

McCormick, Jack, *The Life of the Forest.* Our Living World of Nature Series. New York, McGraw-Hill, 1966.

McIntyre, Joan, ed., *Mind in the Waters; a Book to Celebrate the Consciousness of Whales and Dolphins.* New York, Scribner's, 1974.

Matthews, L. Harrison, *The Life of Mammals,* Vols. 1 and 2. New York, Universe Books, 1970.

Meggers, Betty J., Ayensu, Edward S., and Duckworth, W. Donald, *Tropical Forest Ecosystems in Africa and South America: A Comparative Review.* Washington, D.C., Smithsonian Institution Press, 1973.

Milne, Lorus J., and Milne, Margery, *The Mountains,* rev. ed. Life Nature Library. New York, Time-Life Books, 1970.

Moynihan, Martin, *New World Primates: Adaptive Radiation and the Evolution of Social Behavior, Languages and Intelligence.* Princeton, N.J., Princeton University Press, 1976.

Myers, Norman, *The Long African Day.* New York, Macmillan, 1972.

Reader's Digest Fascinating World of Animals; a Unique "Safari" Through Our Strange and Surprising Animal Kingdom. Pleasantville, N.Y., Reader's Digest, 1971.

Ride, W. D., *A Guide to the Native Mammals of Australia.* New York, Oxford University Press, 1970.

Robinson, H., *Biogeography.* London, Macdonald and Evans, 1972.

Scheffer, Victor B., *A Natural History of Marine Mammals.* New York, Scribner's, 1976.

Stonehouse, Bernard, *Animals of the Arctic: The Ecology of the Far North.* New York, Holt, Rinehart and Winston, 1971.

Van Gelder, Richard G., *Biology of Mammals.* New York, Scribner's, 1969.

Vaughan, Terry A., *Mammalogy.* Philadelphia, W. B. Saunders, 1972.

Walker, Ernest P., and others, *Mammals of the World,* 3rd ed., Vols. 1 and 2. Baltimore, The Johns Hopkins Press, 1975.

Wilson, J. Tuzo, ed., *Continents Adrift; Readings from Scientific American.* San Francisco, W. H. Freeman, 1972.

Index

Related species can be located by referring to the Mammal Classification chart on pages 16–17.
Picture references in italic numbers.